# THE SILVER ECLIPSE

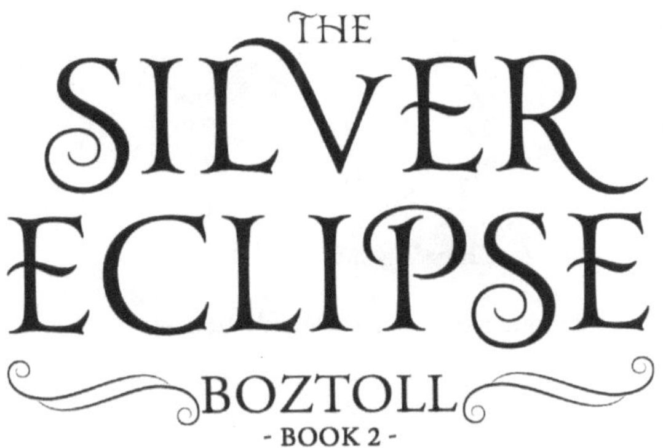

## BOZTOLL

### - BOOK 2 -

## KRISTY DIXON

Boztoll

The Silver Eclipse Book 2

Copyright © 2023 by Kristy Dixon

by Meegore Publishing LLC in The United States of America



Book Cover by Miblart

Edited by Michelle Clark

ISBN 978-1-960841-03-2 Paperback

ISBN 978-1-960841-05-6 Hardback

ISBN 978-1-960841-04-9

First edition 2023

kristydixonbooks.com

For Tom

# CHAPTER 1

Wren sprinted through the jungle as fast as her legs would carry her. The harsh wind whipped against her face and sent her long red hair and cape flying behind her. She jumped as roots broke from the ground, trying to trip her. One scraped her leg, but not enough to slow her. As soon as the roots stopped attacking, the vines overhead lashed at her face. She ran in uneven patterns to avoid their sting.

Another root popped up and Wren jumped, dropping to the ground as a vine missed her face by inches. She rolled and almost landed on a small dragon.

"Ben!" she yelled, as she scooped him into her arms on her way up. She didn't pause, but kept running with Ben tucked safely under her arm. The wind was getting stronger, but she pressed on. A horrible odor filled her nose. She ignored it and kept moving. Focusing ahead, she saw two figures. One was still, with hands raised towards her. The other one was rotating his arms in irregular patterns. Just a little further, and Wren would reach them.

The wind and forest quieted as the two figures lowered their arms, pointing at her, speaking frantically to one another. Wren ran towards them and watched curiously as they tripped over themselves, trying to back away.

"Stop Wren!" her friend Ming Li called when she'd almost reached them. She stopped and wiped her brow, allowing herself to feel the fatigue.

"Ahhh man," Tal said, covering his nose with his cape, "Why are you holding a rednax?"

Ming Li took a few more steps backwards. "You're going to stink forever."

"What are you talking about?" Wren asked, looking at Ben. It wasn't Ben. A reddish brown rednax gazed innocently at her. "Oh boy," she groaned, gently placing it on the ground. It glanced up at her again with its big brown eyes and pushed its snout into her leg. "What do I do?" she asked Tal.

He ran his hand through his wavy brown hair and blew out a slow breath. "There isn't really anything to do now. He already sprayed, so he won't be able to do it again until tomorrow."

Ming Li nervously pulled on the long black braid hanging over her shoulder. "Are you sure?"

"Positive," he said, bending and holding his hand out to the creature. It waddled over and licked it. "Rednax always get a bad rap because of the stink, but they're friendly little things."

"Did he spray me?" Wren asked, sniffing her arm. "Oh, he must have! This is terrible! I can't smell like rednax. We have too much to do! Ugh! My side is wet. This is so gross!"

"What is a rednax exactly?" Ming Li wondered, still keeping a safe distance. "I've heard a lot about them, but this is the first time I've seen one. They kinda look like a skunk pig."

"That's a pretty close description," Tal said, rubbing its head. "They also have venom, but nobody knows why. It's pretty useful to us, but we can't figure out what they use it for. They're herbivores, so they don't need their fangs. I've never heard of anyone getting bitten by one. If it's a defense mechanism, no one seems to know how they use it."

"What am I supposed to do? I can't go into the meetinghouse stinking like this."

"Yeah, the stench is burning my nose. Why did you pick it up?"

"I thought he was Ben. I swear it was green!"

Ming Li snorted, and Tal laughed.

"If you laugh at me, I will hug both of you!" Wren threatened. They both stepped back.

"A jungle rednax is green until it sprays," Tal explained. "They turn that reddish color with the yellow stripe on their tail until they are ready to spray again. The ones you've probably seen around Akkron don't change color."

"It was an impressive move, the way you grabbed him and jumped right back up," Ming Li said. "You've gotten a lot faster, and you almost seemed coordinated out there. I was hitting you pretty hard with that wind."

Wren crossed her arms and glared at her friend. "Gee thanks."

"You have gotten better," Tal agreed. "Sorry about hitting you in the face."

Wren reached up and touched a spot below her eye. She looked at her fingers. There was blood, but it wasn't bad. The last few weeks of training had brought a lot of scrapes and bruises. She could tell she was stronger and faster than when they started.

"It's only a little cut," Ming Li said, taking another step back. "Sorry Wren, I can't take this smell. It's going to haunt my nightmares. I'll go see if Graham has something in his books about getting rid of the odor." She turned and raced through the trees.

Wren sighed and glared at the rednax. It didn't look like Ben at all. The only similarity had been the color. "If they really are friendly, why did he spray me?"

"You're friendly, but you'll attack someone if they threaten you, right? You probably scared him to death. They don't spray on purpose. It only happens when they're afraid." Tal stood and coughed. "Man, my eyes are burning. We might have to wait on our plans if we can't get rid of the odor. The Dark Cloud will know we're coming from a mile away."

She shook her head in exasperation. "We've been waiting forever. We can't push it back again. I'll take a bunch of showers or something." This was so annoying. They couldn't put it off and lose any more time. The longer they let The Dark Cloud run free, the worse the world was going to get. She was going to get rid of this unpleasant smell, even if it meant scrubbing off a layer of skin.

Tal coughed again. He looked to where the rednax was nuzzling his leg. "That's quite the fragrance you have there, buddy."

"You don't need to stay with me. I know it's bad." Her eyes were watering, and it was getting hard to breathe.

"You can't stay out here forever. Just go through the house as quickly as you can and run to the shower."

"There aren't any windows inside to air things out," she said, trying not to gag. "I'll try to wash it off in the river."

"The water's pretty cold, and it's a long walk."

"That's okay. I might be able to run and get away from the stink."

Tal scanned the trees, frowning. "I don't think the river is a good place to hang out by yourself. I don't like to freak you guys out, so I keep most animal activity to myself, but there are some dangerous things around here, especially around the water."

"Can't you tell them to stay away?" she asked, shifting from one foot to the other. It was cool Tal could sense animals, but also unnerving. Sometimes she wondered if it was irritating to have all that going on in his head.

"Well, I can tell them, but that doesn't mean they'll obey. I've noticed the more vicious the animal, the less likely they are to listen. It would be nice if I could communicate with them better."

"I thought you had conversations with them."

He shook his head and sighed. "No, I can sense what they're feeling, but that's not always helpful. I can send them messages, but there are still a lot of things that confuse me. Let's walk down to the river. If I'm with you, we might be able to avoid anything too dangerous."

Wren nodded. She should have known this was going to be a bad day when Brake made breakfast. She could still taste the raw fish scones. That was a taste that didn't die easily.

"Come on," Tal said, walking in the river's direction. "That smell is going to make me puke."

"What are you making?" Ming Li asked as she entered the kitchen and peered into Graham's mixing bowl.

"It's supposed to be a cough medicine," he said, sighing. It was thick enough to choke a horse. This was his third attempt, and it looked like there was going to be a fourth.

"It looks nasty," she said, wrinkling her nose. "You should make it a more palatable color. That looks like snot."

"I don't know what I'm doing wrong," he muttered, scanning the recipe. "The anti-itch cream worked on the first try."

"I'm sure Tal appreciated that. He probably wouldn't have been as willing to help if you made him touch that itchy stuff too many times. He deserves a reward for touching it once. I've never seen skin rash up that fast."

"Yes, but my cream worked so fast he didn't have time to complain." Ming Li was right. Tal deserved something. It was a good friend that let you experiment on them.

"What's wrong with this?"

"I don't know. It doesn't even have that many ingredients!" He ran a frustrated hand over his head.

"Why aren't you cooking it over the stove?"

"I don't see what good that would do."

"Stir it over heat." She pointed to the recipe. "It says right there."

"What?" he exclaimed, picking the book up and rereading it. "How did I miss this?"

"I keep telling you guys. We're overdoing things. It feels like we've been training and preparing non-stop for two solid months."

"We probably should sleep more." He walked over to the garbage can and dumped the goop. Cleaning the bowl of the foul, sticky substance wasn't going to be easy, and he was exhausted. All the late night training and reading was causing them to make mistakes.

"How did Wren's practice go?" he asked, dropping the bowl in the sink and filling it with water. Future Graham could deal with it later.

"Riiiiiight ... that's why I came in. Any chance you have a book that tells you how to get rid of rednax odor?"

"Oh no. Why?" he asked, washing his hands.

"Wren did awesome today. Unfortunately, she accidentally picked up a rednax, and it sprayed her."

"Oh man," he laughed. "How do you accidentally pick up a rednax?"

"She thought it was Ben. She snatched it right up in the middle of an awesome avoidance move. I was impressed, but it smells soooo bad."

"Ben's been in here all morning," he said, pointing under the table. "I keep trying to get him to leave, but he won't."

"It must be hard, being scared of little dragons," she teased, as she squatted and reached out to the small, sleepy creature. He wandered lazily out and let Ming Li pick him up. "You are such a good boy." She scratched his head. "Have you been scaring Graham again?" Ben rubbed his head against her and purred.

"I'm not scared of him anymore," he protested, "But that doesn't mean I want him drooling all over my stuff." He dried his hands on a towel. "So were you out there pelting Wren with wind again?" he asked, gesturing to her windblown braid.

"Of course. Resistance makes you stronger."

"And it's good for your hair," he joked.

She smoothed her braid and glared at him, "Yeah, well, we can't all have your perfect, mess resistant hair."

He smiled and ran his hand over his short black curls. "You could keep yours this short if you wanted. I wake up in the morning and I'm ready. I could give you a haircut."

"I'm good. Maybe you haven't noticed, but every girl in this world has long hair. I stand out enough as it is," she said, undoing her braid. "Do you have any ideas for Wren?"

"I read something about getting rid of rednax odor, but I'll have to find it." He picked up his book and flipped through it. "Sometimes I miss the internet. Everything you ever wanted, right at your fingertips."

"Yeah, but life is more boring when you use screens all the time," she said, taking a seat on a kitchen chair. Ben licked her face, but she didn't seem to mind. "I do miss movies sometimes."

Ming Li had come to this world with her mom when she was nine. She was the only person from Earth besides him. Graham had come here by mistake three months ago when he accidentally followed Wren through a portal. It felt like a lot longer. He could go back home if he wanted, but he felt a lot more useful here. It was also nice to be surrounded by people who cared about him.

Two months ago he had gone with Wren, Ming Li, and Tal to the Island of Meegore, where a crystal had chosen the four of them and allowed them to go into a cave that had been closed for a long time. Inside the cave, they were all given powers other people didn't have anymore ... at least as far as they knew.

Not long after arriving here, Graham realized he had magic. He'd been raised by his aunt and didn't remember his parents. He was pretty sure from recent events that his parents had actually been from this world. Ming Li didn't have any magic until they went inside the cave, but now she was catching up with the others. The four of them were working with a group of adults called The Silver Eclipse to expose and capture a group called The Dark Cloud.

The Dark Cloud was opposed to anyone that didn't have magic. They had been working for years to scare and coerce the magic-less out of the capital city of Akkron and to a place called Boztoll. Not content to leave the people alone after chasing them from the capital, The Dark

Cloud caused a dense cloud cover to come over Boztoll so nothing would grow. Most of the Silver Eclipse speculated that they had lost control and messed up the weather all over the continent. The crops hadn't come in well anywhere, and now everyone was relying on imported crops from the trolls or the other continent.

Graham shook his head. This continent and that continent. People in this world did *not* like to name things. Next time he had a free moment, he was naming the world and the continents. There were only two. It shouldn't be that hard.

"Here it is," he said, pointing at a page. "It looks easy enough. All you need to do is mix some moakberries and rednax venom into some soap and scrub it off."

"And you get the rednax venom, how?" Ming Li asked, cocking her head.

"Uhhh, it doesn't say. Brake must know, since they use it in the prison." Brake was the person who usually took charge of The Silver Eclipse. He let Graham stay at his house when he first came to Akkron and even pretended to be his uncle.

A thump sounded in the other room, and footsteps stomped past the kitchen and down the hallway. The worst smell Graham had ever smelled crept into his nostrils.

"Wren, I'm guessing?" he asked with a grimace.

Ming Li wrinkled her nose. "It has to be."

Tal entered the kitchen and sat next to Ming Li. "Any luck with an anti-stink serum? *Please* say yes."

"I found something," Graham said, handing the book to Tal.

"Easy, peasy."

"Yeah, but how are we going to get rednax venom?"

"Well, there is a rednax sitting outside. We couldn't get him to stop following us. I don't see how you would extract anything. It doesn't sound like a fun thing to experiment with."

"Is Wren in the shower?"

"I assume so," Tal said, with a sparkle in his eyes.

"You think it's funny?" Ming Li asked, playfully punching him in the arm.

Tal smiled, looking unrepentant. "I know, I'm sorry. We were going to go down by the river so she could wash off. On the way there, she tripped and landed in a pile of some kind of animal droppings."

Ming Li covered her mouth and nose with her hands, and a small laugh escaped. "Oh gross! Poor Wren."

"I might have accidentally laughed, and that made her mad, so she ran back here. I doubt she's ever climbed a tree that fast in her life."

Graham shook his head and smiled. The place they were staying was a meetinghouse for The Silver Eclipse. The only way to get in was to climb a gigantic tree and when you got to a certain branch, you fell over it and into the house. No one could see the house from the outside.

"Can't I leave you four for a few hours without coming back to a problem?" Brake asked, entering the kitchen. "The whole place reeks of rednax."

"We're working on it," Graham said, holding up his book.

"Where's Wren?" he asked, scratching the black stubble on his chin.

"Shower."

"A shower will do absolutely nothing. There's some rednax venom in the cupboard up there," he said, pointing. Tal jumped up and pulled a stool over to the spot and climbed up.

"This?" he asked, holding out a bottle with something clear inside.

Brake nodded, taking it from Tal and handing it to Graham. "We usually keep a rednax scrub on hand, but we didn't replace it after Austra's last encounter."

Graham wasn't sure, but it looked like Brake was hiding a smile. "Austra got sprayed?"

"Yes. It would be good to not mention it to her. I don't know if you've noticed, but she doesn't have a sense of humor."

Ming Li rolled her eyes. "I think we noticed a few times."

"I'll go pick some moakberries," Tal offered. "I saw some not too far away."

"I bet you just want to get away from the stink," she accused. He smiled and walked out the door.

Brake shook his head. "We may need to bring Brog back to babysit."

"He wouldn't have been able to stop the rednax," Ming Li said with a grin. "And Brog might be big and intimidating, but once you get to know him, he's a softy. That's why all the animals like him."

"That's true," Brake admitted, "But he has excellent hearing, which is convenient when you all make danger-ous, or underthought plans."

"He only told on us a couple of times, though," Graham said, searching the cupboard for a clean bowl. "The other times he just shook his head."

"What other times?" Brake asked, his eyebrows knitted together. "Is there anything you need to tell me?"

"Nope," Ming Li said. "I'm pretty sure you've already found out every crazy thing we've done."

"I doubt it, but I'm feeling too tired for extra surprises."

# CHAPTER 2

"Why is that thing in here?" Wren asked, pointing at the rednax sitting on Tal's lap. She was already in a bad mood, waiting for Hedder and Tram to show up. Brake, Zera, Brog, Hamble, and Austra were already there, relaxing on their colorful chairs, eating cookies.

"He can't spray you," Tal told her, giving him an affectionate rub behind the ears. "So I'm not sure why it matters. He's lonely."

"We can't collect pets," Brake said, raising an eyebrow at the pampered rednax. "That's why we sent Walter to stay with Brog."

Walter was Graham's dog. He was a happy, bouncy goldendoodle with way too much energy for the meetinghouse. Ben was only allowed to stay because he didn't need much attention and he was litter box trained. Wren liked animals, but she wasn't thrilled with the rednax at the moment. She looked at her red arms. It had taken a lot of scrubbing to get rid of the stench.

"I'll put him outside as soon as he calms down," Tal said, patting the rednax. "Wren terrified him, scooping him up and darting through the jungle like a madwoman."

Wren shot Tal an irritated look. It figured he was taking a stinky animal's side over hers. She knew she was being unfair, but she was tired, hungry, and ready to be doing something. Two months in this place was grating on her nerves. Austra was sitting as far from Tal as she could without leaving the room. She was sending the rednax the same hateful stares Wren felt.

"Hedder and Tram just messaged and they can't come, so I suppose we can start," Brake said. "I heard from Drew today."

Wren sat up straight. Drew was her father. Akkron's governor had falsely accused him of belonging to The Dark Cloud and had him arrested. Drew escaped and was now in hiding.

"Is he alright?" she asked.

"He is. He traveled to the other continent to hide. Governor Briggs has no authority there."

"Don't look at me like that," Tal said, catching Wren's glare. "I'm not my dad."

Wren blinked. "I know, sorry." Tal's father was Governor Briggs. Tal was nothing like his dad, but it sometimes made things awkward.

"Drew doesn't have any sense," Austra said, in her stilted accent. "He probably thinks he's safe on the other continent and he won't be careful."

"He will be fine," Zera cut in before Wren could react. "Drew is far more capable than some people think. Now why are we here?"

"We're ready to go to Boztoll," Graham said resolutely. "It's time. We've been training and we want to see what it's like there."

Austra sighed. "So you want to go on a vacation? I told you working with children wouldn't get us anywhere."

"No, we want to see what we're up against," Ming Li said, putting her hands on her hips. "We want to scope it out and see if The Dark Cloud is there, and see if we can stop the clouds."

"So you think you can just walk in and figure out what the rest of us can't?"

Ming Li shrugged. "Maybe."

Brog raised his hand. "If I may speak?" Even sitting on his plush green cushion, he remained eye level with those on chairs. "Haltina would like to speak to Wren."

She jolted in her seat. "Why me?" Wren had only talked to Haltina for a few minutes when they were on the Island of Meegore. Haltina didn't know Wren anymore than she knew the others.

"She has some things to say to you she believes will be of use."

"We could go after we go to Boztoll."

"She would like you to come today."

"It isn't wise to disobey a giant," Hamble said, rubbing his bald head.

"What if Wren opens a portal for us near Boztoll, and she can join us when she's done?" Graham suggested. "It probably won't take too long, and we can start watching Boztoll until she comes."

"Watching Boztoll?" Austra scoffed. "I hope your plan isn't to hide in the bushes and see if you witness anything interesting."

Tal, Graham, Ming Li, and Wren all glanced at each other. That was exactly what they were planning on doing.

"We can't send Wren on her own," Ming Li said, shaking her head. "Last time we did that, she got caught by Professor Dovin."

"I can go with her," Austra volunteered. "It's been a long time since I was in Meegore."

"I can come too," Hamble said.

"Alright, so Wren is taken care of. I don't know if I want the three of you going without supervision."

"I have a meeting with the Fleet," Zera said, apologetically. Zera was part of the Governor's Fleet, like Wren's dad had been. She was one of seven members that counseled the governor.

"The less of us, the better," Tal said. "We don't want to draw attention. We can send a message if we need someone."

"Alright," Brake agreed. "Let's all meet back here in two days."

Austra shook her head. "Alright? You all let these children have too much freedom."

"All we're doing is looking around," Graham said. "What trouble could come from that?"

Graham, Tal, and Ming Li squatted behind a large boulder and stared at Boztoll. It surprised Graham they called it a city. He would call it a village. A rundown village. The houses were considerably smaller than any he'd seen in Akkron, and they were weathered with their peeling, dull paint, and saggy roofs.

"Are all the houses like this?" he whispered.

"No," Tal assured him. "I thought we should come to this side of town because bad things are more likely to happen in the rundown part of a city."

Ming Li wrapped her black cloak tightly around herself. "It's crazy how dark it is. We need to figure out how to get rid of those clouds. It's only a little past noon." They watched people walk up and down the cobbled streets, going about their days like the darkness wasn't anything unusual.

"There has to be an easier way to summon the silver eclipse," Graham said, scanning the dark sky. "It would all be a lot easier if we weren't always hiding. It's hard to figure anything out when we can't talk to anyone."

"This is super boring," Ming Li complained, kneeling on the ground. "Aren't you all glad I had you wear your jeans? They are so much more practical when you have to get down and dirty."

"Yeah, but if anyone sees us, they'll wonder what we're wearing." Tal said, looking at his pants. They all had some Earth clothing from the time they had gone to talk to

Graham's aunt ... well, the person he had always thought was his aunt. It turned out she was just someone who had begrudgingly taken him in when his parents disappeared.

"You're just upset we didn't let you wear your baby shark pajamas," she snickered.

"Hey, those things are comfortable. I don't get why you guys are always making fun of them."

"They are slightly better than your flamingo swimsuit."

"Someday, I'm actually going to go swimming in those."

"There isn't much going on here," Graham said, glancing around. "We should go around the perimeter and see if there's more going on somewhere else."

Ming Li made a high-pitched squeak, causing Graham to turn. She was gone. He was about to alert Tal when something smashed into the back of his head.

Graham had gone for almost sixteen years without ever getting knocked out, and now it had happened twice in only a few months. This better not be his new normal. He groaned and tried to move his arms, but realized they were tied above his head. At least he'd been tied in a sitting position. He squinted through the dim light. He appeared to be in a small, unfurnished cabin. The floor was hard packed dirt, and the walls were crudely trimmed logs.

"Oh my head," Ming Li mumbled, from four feet away from him. Her hands were also tied above her head and being held up with a rope that was fastened to the wall.

Past her was Tal. His chin was resting close to his chest. He must still be out.

"Are you alright?" Graham asked.

"I will be," she muttered angrily. "You have a bloody bandage on the side of your head."

Graham wished he could put a hand to his head to see how bad it was. Ming Li looked alright. She wasn't bleeding. Tal didn't appear to be either.

"Tal?" he said loudly. "Tal, wake up!"

"Hmm?" His head came up a bit and rolled to the side.

"Talon! Wake up before I kick you!" Ming Li threatened.

"You don't need to threaten him."

"If he doesn't wake up, it'll be harder to escape."

"I escaped the jungle cows," Tal mumbled, lifting his head.

"Great, what if he has a concussion?" she asked, struggling with her ropes.

"What's going on?" Tal asked, shaking his head. "Where are we?"

"A cabin or something," Graham told him, pulling on the ropes. They didn't budge. "I don't know how we're going to get out of this. Did anyone see anything?"

"No," Ming Li said. "The last thing I remember is staring at Boztoll."

"How come I'm the only one bleeding?"

"I have a headache, but I don't think they hit me."

"I remember someone covering my mouth with something," Tal said, through squinted eyes. "It must have been ailam powder. It smelled like sunshine."

"What the heck does sunshine smell like?" Ming Li asked, shaking her head and looking at the ceiling. "This is a rude way to tie someone up," she said, awkwardly trying to stand. All she could manage was a squat. The rope was anchored too low to allow anything else. She sat back on the floor. "This is stupid."

"These knots are really tight," Graham said, struggling against his bonds. "Can you untie us?"

"No," she said. "They tied my hands funny, and I can't seem to do any magic."

Tal chuckled. "Does anyone else see floating pill bugs? I didn't realize they could float."

"Great," Ming Li said, rolling her eyes. "It appears Tal can't handle ailam powder."

"No guys," Tal smiled. "This is a good thing. If they leave us here, we can eat the bugs. They might float into our mouths." He closed his eyes and opened his mouth.

Graham and Ming Li shared a glance and went back to struggling against the ropes. The door across the room opened, and a figure entered. Even though it wasn't bright outside, the small amount of light from the door still caused Graham to recoil.

"Oh good, you're awake," the figure said, shutting the door behind him. Graham stared in surprise. He was expecting Professor Dovin, not a kid his age. He had a young but solemn face and he was wearing dirty clothes. His straight black hair hung limply to the collar of his tunic.

Ming Li's mouth turned down. "Don't tell me *you* captured us."

"Why is that a surprise?" the boy asked, squatting in front of her.

"Well, because you look like the wind might knock you over. You could seriously use a few cheeseburgers."

"I had friends helping me," he said, glaring at her. "You didn't give us any trouble at all."

"Are they as strong as you? That would explain why you had to attack us from behind."

"I don't have to explain myself to you. You're the ones tied up." He glanced over at Tal and frowned. "What is your friend doing?"

"Tal, knock it off," Ming Li said.

Tal opened his eyes and closed his mouth. "My name is Talon," he said, using an unnaturally deep voice. "Don't call me Tal."

"Why did you bring us here?" Graham asked the boy.

"As if you didn't know," he scoffed, standing. "You want to sneak in and cause problems? You aren't very good at hiding. I spotted you as soon as you jumped out of that portal. We won't let you ruin what we've started here."

"We have friends that know we're here," Graham said, keeping eye contact. He forced himself to remain calm. "They'll come for us."

"Hopefully, they're more skilled than you."

"My dad is the governor," Tal said, glaring at the boy, "And he will come here and make all the people eat bread and clean the livestock."

The boy's eyes narrowed. "What are you talking about?"

Ming Li glared. "He's having a reaction to your ailam powder." Graham was sure it was eating her up inside, not being able to put her hands on her hips. "The least you could do is get him some water or something to help."

The boy seemed uncertain for a moment. "I can bring him water and perhaps a tart."

"Your brother eats tarts!" Tal said, laughing.

"I'll be back," he said, side-eying Tal. He hurried out the door.

"I can't believe that guy captured us," Ming Li muttered.

"I can't believe I'm not running on the beach with Wren, riding unicorns and eating pill bugs!" Tal yelled. He looked angry and then started to cry.

"This is just sad."

"You can't judge someone when they are out of their mind," Graham said, trying to move his hands. If he could loosen the rope a little, he might be able to do magic. He tried sending a whispered message to Brake, but it wasn't working.

The boy came back in with an older man. The man had a black mustache and was wearing something that resembled a leafy green sombrero. It seemed out of place with his dark cloak. They both glanced to where Tal was crying on the floor.

"Give him the water," the man said. The boy walked over to Tal and held a cup to his mouth. Tal took a couple of mouthfuls and spit it in the boy's face. The boy growled and wiped it off on his dirty sleeve.

"Hey, I'm a whale!" Tal said, sniffing. "I've always wanted to be a whale."

"Sendo, go get the ailam powder," the man said.

"That wouldn't be a good idea," Graham argued. "He's having a weird reaction to the powder. You might kill him if you give him more."

"And why would we care?" the man asked. "We will probably kill you all, anyway."

"Don't worry papá," Sendo told the man. "I'll take care of them. You need to search for others."

The man nodded. "I don't know how you keep roping me into your schemes. This is the last time, mi hijo." He slammed the door as he left.

"Now I can believe that guy captured us," Ming Li said to Graham. "He's a lot stronger looking than this guy. And he has the evil guy mustache that Tal can't appreciate in his condition."

"I'm strong enough," Sendo said, crossing his arms. "I carried you here."

"You carried me? I hope I drooled on you."

"You did," he said with a smirk.

"I don't want to marry the princess," Tal mumbled. "Wren is cuter and betterer than she is. Haha! Betterer isn't a word. My dad doesn't like words that aren't words."

"If you let me touch him, I can try to heal him," Graham said.

"I don't think so. I'm not falling for something that dumb. Besides, we rubbed rednax venom on you, so the only one who can do magic here is me."

"Give him more water," Ming Li commanded.

Sendo shook his head. "Why? So he can spit it on me again?"

"Talon. You need to drink the water. Don't spit."

"Okay, Li Ming Li Li. No spitting. Has anyone ever told you your hair is very shiny and beautiful?" He asked with a big, toothy grin.

"Just give him some water," she muttered.

Sendo bent down and put the water up to Tal's lips. He sipped it and leaned against the wall, closing his eyes. "I'm gonna take a nap now."

Ming Li rolled her eyes. "Please do."

Graham looked around for anything that could be a weapon, but the place was bare. If they didn't get untied soon, their arms were going to be useless. He was already in a lot of pain and his hand was going numb. When Wren finished at Meegore, she would try to find them. He wished he could send her a warning. It wouldn't do them any good if she came and was captured.

"I'll let your arms down if you don't try anything," Sendo said, pulling a knife from his boot. "Don't get too excited. Remember, you can't do magic."

He cut the rope that was holding Ming Li's arms up, but left the one binding her hands. She slowly lowered her arms and tried rolling her shoulders. She cringed. Graham wondered how long they'd been tied up. Sendo cut his rope, and he lowered his arms. They stung, but it still felt good.

"Don't even think about doing whatever you're thinking about," Sendo said, pointing his knife at Ming Li. "Even if you take me out, there are guards outside, and they won't have mercy on you."

"What?" Ming Li shrugged. "I wasn't planning on doing anything. I mean, I thought of kicking you in the backside, but I wouldn't want to get my boots dirty. Have you ever heard of bathing?"

He glared at her as he cut Tal free and lowered him to the floor. Tal snored and didn't notice.

"It would be easier to take a bath if I wasn't always on the lookout for people like you."

"You don't know anything about people like me."

"I know more than you believe, and if you think you can stop us, you're wrong."

"Where's Dovin?" Graham asked.

Sendo's eyes widened. "I don't know anyone named that."

"Then why do you look guilty?"

"I can look however I want. I'm the one with the magic and the knife."

Graham couldn't figure it out. It seemed sloppy of The Dark Cloud to leave these people in charge. He'd expected more intimidating Karlof looking characters to be running things here. Wren's uncle was the scariest looking person he'd seen since coming here.

"You're going to have to give up eventually," Ming Li told him. "People in Akkron are getting angry. You can't take on an entire city."

Graham wanted to stand, but he wasn't sure what Sendo would do with the knife.

"We'll fight anyone who opposes us. We have a worthy cause, and so we must succeed," Sendo growled.

"Whatever," Ming Li said, shaking her head. "And what are you planning on doing with us?"

"The same thing we do with any of your people that come here."

"And that would be?"

"You don't want to know."

Ming Li tilted her head. "I actually do want to know."

"I guess you will have to wait and see."

# CHAPTER 3

"It's nice to feel the sun on my head again," Hamble said, rubbing his pale, bald head. "The Dark Cloud must not be bothered with the weather in Meegore."

Wren scanned the island, hoping to see Haltina. "It's been warm in the jungle, except for those few ice storms we had a few weeks ago. They didn't last very long, though."

"Akkron's weather remained unchanged until recently," Austra said, pushing her light blue hood off her head. "It's too bad it all turned cold before they harvested the crops."

Wren studied the pretty olive skinned woman and shook her head. Austra was wearing a pink dress that flowed past her knees. The blue cloak and pink dress reminded Wren of a birthday cake Ming Li's mom had made for her a few years ago.

"Your dress is going to get dirty," Wren warned her.

"Oh, I know," Austra said, looking Wren up and down, "But that's no reason to dress gloomy."

Wren wouldn't let Austra annoy her. She'd copied her friends and wore her Earth clothing. The pants were a lot thicker than what she usually wore, and if she ended up going into the cave, she wanted to be prepared. Her face burned as she remembered how she'd worn a hole in the backside of her pants the last time she was here.

"Where is Haltina?" All Wren could see were trees and purple crystals. "I didn't see any houses last time we were here."

"And you won't," Austra said. "Giants don't live in houses. They usually avoid any structure that isn't natural."

"So they live outside?"

"Yes. They don't like to mess with nature. They enjoy the world being untouched by anything modern."

"I haven't been here in a long time," Hamble said, as they walked down a dirt path, through the dense forest. "The trees have grown a lot. The crystals have as well."

They came to an opening and saw the large crystal that determined who could enter the cave at Meegore. Behind it was a vast cave that looked more like a big hill when it wasn't open. Wren could still remember the feeling of disbelief that had run through her body when the crystal had lit up for her and her friends. The day they realized they had been chosen to help save the world.

"That's an enormous crystal," Hamble said. "I never took time to appreciate it the last time I was here. I was too distracted."

"By what?" Wren asked.

He shrugged. "Work."

"What were you working on?"

"I don't know how much Brake wants you to know."

"Seriously Hamble," Austra said, rolling her eyes. "You are as bad as Drew at keeping your mouth shut."

Wren was going to question him more when two giants came out beside the cave.

"Oh, it's you," one said, bowing to Wren. "We wondered if you would be back. I am Tnarg, if you don't remember. The T is silent."

"Yes, I remember you," Wren said, trying not to smile. She should start introducing herself as Wren. The W is silent. "And you are Mot," she said to the other one.

"She remembers us," Mot said, smiling. His long hair was pulled back in a braid, and he wasn't wearing any shoes. Wren hadn't met a giant yet who wore shoes. Brog could outrun them all barefoot, so she wasn't going to criticize it. They both wore swords at their sides.

"Are your friends here as well?" Tnarg asked.

"No. I came to talk to Haltina."

"Thank you for coming," Haltina said, emerging from the trees. Wren turned and studied her. Brog had told them she was one of the oldest giants alive. She wondered how old the woman was. She had a lot of wrinkles and her hair was silver, except for one streak of brown woven into her braid.

"I hope you don't mind Hamble and I accompanying Wren," Austra said, bowing her head to Haltina.

"Not at all Austra. It is good to see both of you again. It has been a long time," she said in a feeble voice. "Who are you chasing these days?"

"Uh, just The Dark Cloud," Hamble said, glancing sideways at Wren.

"I remember when you were all young, running around, trying to save the world. It was hard to have a conversation without hearing of your escapades."

"Yes, well, now we choose to work more discreetly," Austra said, clasping her hands together.

"As you should. And you are training up the next generation."

Hamble looked at the ground. "Something like that."

Haltina motioned to some fallen logs that had been placed in a square. "Come sit." She led them over and sat awkwardly on a log. It creaked under her weight.

Wren sat across from her and tried to imagine Hamble and Austra running around trying to save the world. Her lip twitched when she tried to picture Hamble as a young man with hair.

"Is something amusing you?" Austra asked, as she sat gracefully on the log next to her. She didn't seem worried about snagging her dress.

"Sorry. It's strange to have someone talk about you guys running around hunting bad guys."

Austra sniffed. "We still do that."

"They weren't a group you wanted to mess with," Haltina laughed. "When Governor Jorin assigned them all to protect Akkron, they took their job seriously. They became a symbol of hope for everyone. It was my job to help counsel them, although they didn't always listen."

"Just the two of them?"

"Oh no. Let me think … it was Austra, Hamble, Fria, Drew, Brake, Tram, Zera, and Magnalee."

Wren's eyes widened. "My parents were part of it?"

Hamble cleared his throat and Austra pursed her lips into a tight line.

"Yes. They never told you?" Haltina asked in surprise. "It was an experiment the former governor had. The guard was failing to keep the peace, so he chose those he believed would be good at upholding the law. Their first assignment was to bring in a dangerous group of criminals at all costs. They weren't bound by any of the laws. The governor gave them permission to do anything they needed to."

"It worked well in the beginning," Austra added, staring blankly at the dirt in front of her. "It had the potential of being extremely efficient, but everything went wrong."

Hamble nodded. "Once we began following criminals to Earth, it all fell apart. That was when we first became aware of The Dark Cloud."

"I always thought going through portals was this huge crime, but you were all doing it," Wren said, scrunching her forehead. "And how did you control which world you went to?"

"We were buying portal dust illegally," Hamble admitted, focusing on his hands. "The goblins were selling it to us. They had one type of dust to go to Earth, and one to come here. We ran out at the worst time."

Wren wasn't sure, but it looked like Austra quickly wiped a tear from her eye.

"My dad traveled to Earth?"

"No," Hamble said. "The only ones who ever went were Brake, Zera, Tram, Austra, and me. Magnalee, Drew, and Fria were keeping track of things in Akkron."

"What happened?"

Hamble shrugged. "It's not our story to tell."

Haltina tilted her head and sighed. "There are always so many secrets. Do they help?"

Austra took a shaky breath and shook her head. "We brought back most of the people we were after. There was only one left."

"The one who hurt Brake?" Wren asked, remembering Ming Li's story of coming to Akkron. Ming Li and her mother Mali had found Brake injured and tried to help him. He ended up bringing them through a portal to Akkron.

"Yes," Austra muttered. "Years later Brake was finally able to get his hands on some more dust. He returned to Earth to find the man ... among other things. When he was there, he got injured. It wasn't his plan to come back to Akkron yet. He had something important to do. Because he was injured, he wasn't thinking clearly, and he came back for the last time. He paid dearly for that decision."

"That last week we were together was the worst week of my life," Hamble said, tightly crossing his arms. "We didn't catch the man. We lost Magnalee, and Brake lost his family."

Wren swallowed a lump in her throat as she thought of her mother. She never knew what had happened to her exactly. She only knew she'd been working to get rid of The

Dark Cloud. And Brake had a family? She had so many questions, but she was unsure what to ask.

"What happened to my mom?" she finally asked. "My dad doesn't talk about her at all. All he ever told me is that she died trying to stop The Dark Cloud." She glanced from Austra to Hamble. Austra was pulling at a loose thread on her dress and Hamble shook his head.

"Fria started acting strange," Haltina said, when the other two didn't answer. "We later found out she had made a deal with The Dark Cloud. Magnalee found something out and confronted Fria. Fria pushed her off of a cliff when they were waiting for the rest of the group to show up. Magnalee never could levitate herself. Zera saw it all. Fria noticed Zera and panicked. She jumped into a portal and we never saw her again."

"So she's still running free out there somewhere?" Wren asked, covering her face with her hands. She peeked over her fingers. "Did you try to find her?"

"We did try," Austra said, softly. "We are still trying. It's been over fourteen years though, so the chances of ever finding her are slim. Everything fell apart at once and Akkron turned on us. They saw Fria's betrayal as a reason to stop supporting us. The governor agreed. Our group broke up, and we tried to find separate lives away from each other ... at least that's what we wanted everyone to think."

Hamble scratched his head. "We all split up and didn't come back together until Brake had the idea to build the meetinghouse. We had as little contact with each other as we could. Of course, Zera and Drew both ended up running for Fleet positions. We thought it would be good to have someone close to the governor.”

“We never imagined they would both be elected,” Austra said. “Especially after people saw our last mission as such a failure, which it was. It’s amazing they have both been able to stay on the Fleet for all these years.”

“I’m pretty sure my dad has been kicked out.” Wren’s brain hurt. She wished she could remember her mom, but she’d been a baby when she died. Her dad wouldn’t talk about it.

Hamble sighed. “Now we work in secret and hope no one finds out.”

“If people knew what you were doing, why is everyone so blinded about The Dark Cloud?”

“We didn’t call them The Dark Cloud back then,” Hamble explained. “People only knew we were chasing criminals. There was a lot of gossip.”

Haltina reached across and put a wrinkled hand to Wren’s face. “This is not why I brought you here. Now I can see that you are troubled. I would suggest the three of you stay here for the night, and we can talk tomorrow. Spend the rest of the day walking around the island and breathing in the clean air. I don’t think you will be able to focus on what I have to say now.”

Wren nodded. She rubbed her eyes and held back tears. Whatever Haltina wanted to tell her would wait. She would have to catch up with her friends later than she had planned. She hoped they wouldn’t mind.

"Where are we?" Tal moaned, rubbing a hand over his eyes. He sat up and inspected the cabin.

Graham was glad he'd finally woken up. He was starting to worry. Sendo had untied their hands so they could sleep last night, and Tal hadn't even moved. He'd slept all night while Ming Li and Graham had tossed and turned on the floor, after their dinner of bread and water.

Ming Li yawned. "It's about time you woke up. We've been captured by The Dark Cloud."

Tal shook his head and frowned. "Really? I thought we were being careful."

"They caught us from behind," Graham said. "I guess they saw us come out of the portal. We didn't have a chance."

"Wait, I'm remembering something … was there a guy in a really big hat?"

"Yes."

He grimaced. "It's coming back. I'm hoping most of what I'm remembering was a dream."

"You can keep hoping," Ming Li said, standing up and stretching. "Ailam powder is not your friend."

"Oh man," he said, hiding his head on his knees. "It's all a little blurry, but I'm pretty sure I was out of my head."

"Or it's possible you were confessing your secrets. Something about running on the beach with Wren, if I remember right," Ming Li teased.

"You can't be held responsible for something you say when you're drugged," Graham reassured him. He noticed Ming Li didn't bring up what Tal said about her shiny hair.

"We aren't tied up," he said, running a hand through his messy brown waves. "Why are we staying here?"

"There are a bunch of people standing guard all around the cabin. They put rednax venom on our arms so we can't do magic. We tried rubbing it off, but that stuff is like glue."

Tal sighed. "How many people are guarding us?"

Ming Li shrugged. "For all we know, there isn't anyone out there and we're staying in here like a bunch of idiots."

"There was one boy around our age and his dad," Graham said. "They said there are people outside guarding. They can all use magic and we can't, so we haven't decided what to do."

Ming Li pulled her black hair over her shoulder and started braiding it. "I say we do what we did with Karlof. If anyone comes in, we surprise them and take them without magic."

Graham tried to think. Taking out Sendo should be easier than Karlof. Karlof was Wren's uncle who had tried to force Wren to take him to Meegore to get magic. He was huge, but the four of them had managed to knock him out. He wasn't sure, but he thought in the end they probably looked more beat up than he did.

"That might work, but what if his dad is with him?" Graham asked. "That guy was pretty big. And if there are other people guarding us, they could run in and take us out easily if we can't use magic."

"Shhhh ..." Tal said, leaning towards the door.

Muffled voices were coming from the other side. Graham strained his ears, but he couldn't make out what they were saying. The door opened and Sendo came in, followed by two boys who looked like they might be his brothers. One was taller than Sendo with wider shoulders, but they had similar shaped faces and eyes. The other one was shorter and a lot rounder. One thing they all had in common was that they were all holding knives.

Tal and Graham both stood, and Ming Li moved closer to them.

"Do we get a last request?" Ming Li asked, eyeing the knives.

"Why would we give you that?" Sendo asked, rubbing his finger along the edge of his knife.

"Come on Sen," the taller boy said. "The girl is muy bonita. Let her make a request."

Sendo narrowed his eyes and sighed. "Fine. If it's something fast."

Graham hoped Ming Li or Tal had something to say to distract them, because he had nothing.

"I refuse to be killed unless someone kisses me," she blurted out. Sendo and his two companions looked startled for a minute before bursting into laughter.

"Seriously?" Tal muttered. "That's the best you can come up with?"

"I have been waiting fifteen years to finally get kissed, so I think it's a reasonable request," she said, putting her hands on her hips and glaring at Tal.

"Granted," the taller boy said. "We will even let you choose who does it."

"Nah, let them decide," Ming Li said, peering at Tal and Graham.

Graham's stomach dropped. He glanced at Tal and saw the same panicked expression he could feel on his own face. The only girl Graham had ever kissed had been Olive Jensen in seventh grade and that was because of a dare. It had gone so bad, that everyone at school had labeled it the ugliest kiss they had ever seen. Olive told everyone he was the worst kisser ever, and he had been traumatized since at the thought of kissing anyone.

"Can we have a moment?" Tal said, dragging Graham to the far corner. They turned their backs on the others and Tal whispered, "It has to be you."

"Me? Why me?" Graham whispered back. "You're the one who flirts with her all the time."

"What? I do not. Teasing isn't flirting."

"It is."

"It isn't."

"Why me?"

"I can't ... just kiss anyone."

Graham rolled his eyes. "And that means I can?"

"No seriously. I don't want to get into this right now. Just do it. As my friend."

"I can't kiss Ming Li. It would be weird."

Tal wiped his sweaty brow. "How many girls have you kissed?"

"How many have you?"

"I asked you first."

"Why is that the deciding factor?"

"Because whoever has kissed the most should do it. My guess is that's you." Tal whispered, sounding panicked.

"It's probably you. You were the one walking around school like, 'Look at me, I'm so awesome.' How many of your fan club did you kiss?"

"Ugh, don't make me say it. I'll owe you forever if you do it."

"If we're going to die, I don't think that helps."

"We aren't going to die," Tal whispered angrily. "The knives are to scare us into telling them something."

"I know," Graham admitted, "So I don't want to be the one walking around trying not to act weird around Ming Li."

"I've never kissed anyone," Tal finally said, shrugging. "I have an arranged marriage, and I promised I wouldn't."

Graham snorted. "An arranged marriage?"

"Yes."

"Who has arranged marriages anymore?"

"Lots of people."

"You don't take me as the type of person to go along with something like that."

"It's not like I have a choice. I've been trying to figure out how to get out of it. I figure once I meet her I'll be so annoying, she'll beg her parents to break it."

"You haven't even met her?"

"No, and I still have five years to figure out how to get out of it."

"I think you're lying, so I'll have to kiss her."

"I'm not."

"You have been whispering for quite some time over there," Sendo said. "Can we please hurry?" They turned their heads to look at him.

"You better be arguing over who gets to kiss me," Ming Li said, crossing her arms.

"Come on!" Tal pleaded when they put their heads back together. "What if we do your game? Paper, scissors, rock, or whatever."

"Fine," Graham mumbled.

"Times up," Sendo said. They turned around and watched him put his knife in his boot. "I'm assuming you're using your time to figure out how to escape. How hard is it to make a little decision?" He walked confidently over to Ming Li and put his hands on her upper arms.

"You better watch it," Ming Li said, narrowing her eyes.

Sendo shrugged. "Watch what?" He leaned in and kissed Ming Li. Her hands fisted, and Graham waited for her to punch him in the head. She didn't. Sendo's two companions whooped, and Tal coughed behind his hand, as the kiss went on longer than seemed right for the situation.

Sendo finally pulled away. "See? Not that hard." He walked a few steps back and took his knife back out of his boot.

Ming Li looked stunned. She sat back against the wall. Graham wasn't sure, but he thought Sendo's ears and neck looked a little more pink than before.

"Good job, hermano," the tall boy said, patting him on the back. "I didn't know you had it in you."

"Callate, Miguel."

"That was awesome," the hefty boy said, smiling. "You're my new hero, Sen! Miguel hasn't ever even kissed a girl."

"Be quiet Juan," Miguel said. "I've kissed lots of girls. I just don't tell you about it."

"This is ridiculous," Sendo growled. "We will come back later." He turned and stormed out of the cabin. Miguel and Juan followed after him.

When the door slammed, Tal chuckled softly. "I sure didn't see that coming."

"You alright Ming Li?" Graham asked. "Sorry we let that happen."

"It's alright," Ming Li said, shaking her head. "His germs were not as gross as they should have been."

# CHAPTER 4

"Have these guys ever said they were part of The Dark Cloud?" Tal asked, trying to peek out of the one small window the cabin had. It was high and covered in a layer of dirt. Graham had already tried looking through it earlier, with little success.

"No," he said, trying to rub more rednax venom off his arm. "We just assumed."

"They don't seem like people they would have recruited."

Ming Li rolled her eyes. "And they're from Earth. I used to think I was pretty cool being the only kid from another world, but now it seems like every other person I meet is from Earth."

Graham smiled at her exaggeration. "You mean me and those guys?"

"Well, I haven't met that many people lately," she said, tossing her braid over her shoulder.

Tal turned from the window. "What makes you think they're from Earth?"

"They keep throwing in Spanish words, and from what I've seen, no one in Akkron speaks Spanish. I don't even think they have magic. If they did, why the knives?"

"So we might be sitting in here for no reason?" Tal questioned, rattling the doorknob.

Ming Li tilted her head to the side and smiled. "Do you think we can ask them for another last request?"

"NO!" Graham and Tal said together.

She laughed. "You're both wimps."

The door opened and Sendo entered, followed by a woman. She was wearing all black except a gold belt. Everything about her looked perfect. Her blonde ringlets were tied back into a neat ponytail, and her fingernails were long and red. She was short and petite. If she was going to try to intimidate them, she was going to fail miserably.

"Talon?" the woman spoke, putting her hands on her hips.

Graham looked at Tal. He was standing there looking as if he had seen a ghost. A scary one.

"What in the world are you doing here?" she asked, walking over to Tal. She reached up and put her hands against Tal's cheeks. "You're filthy. You've gotten so tall! I feel short standing next to you."

"Are you helping these people?" Tal asked, taking a step back.

"They're helping me," she smiled. "I'm in charge around here ... well, at least with this group." Her smile fell. "They brought me here because they captured three Dark Cloud members."

"We aren't Dark Cloud members," Ming Li said, crossing her arms. "We thought they were."

"Then why were you spying on Boztoll?" Sendo accused. "Only Dark Cloud members sneak around."

"We were trying to see what was going on here," Tal protested. "We want to help."

"Does your father know you're here?" the woman asked.

"Nope," Tal said smugly, "And I'm planning on keeping it that way."

"You should be at school."

"Umm ... if you checked in more than twice a year, you might know that school is the last place I can be right now."

"Were you suspended?" she asked, frowning.

"Wow. News doesn't reach Boztoll very fast, does it?" Ming Li said to Graham.

"I guess not," he agreed.

"What news?" she asked.

"It doesn't matter," Tal said. "Can we go now? We haven't eaten in a while."

"Did you run away? I don't understand why you would quit school when you were doing so well."

"I can't go to school, Mom. People are looking for us."

"Why?"

"Mom?" Ming Li and Sendo both asked together.

"Just tell her about The Silver Eclipse," Graham told Tal. Tal had said his mom was nice. Might as well get it all out.

"You know about The Silver Eclipse?" she asked with wide eyes.

"Know about it? We are it," Graham said.

"You can't be part of The Silver Eclipse," Sendo said, looking suspicious. "We are part of it, right Valeena?"

"We named it," Ming Li said, glaring at Sendo. "Well, I guess Graham named it."

A light seemed to turn on in Valeena's eyes. "It can't be. Were you the ones that went into the cave at Meegore?"

Ming Li nodded. "You better believe it."

"Tal? You went into the cave?"

Tal looked around and shrugged. "Sure."

Valeena put a hand to her chest and tears were running down her smooth cheeks. She grabbed Tal and hugged him. Tal's eyes bugged out as he awkwardly hugged her back. Over her head, he mouthed, "Help me!" to Graham.

Graham smiled and shook his head.

"He might need to breathe," Ming Li said. "I'm starving. Now that we know we are on the same team, can you take me somewhere to find food?"

"You can all come to my house and I'll feed you," Valeena said, releasing Tal and wiping her eyes.

"Your house?" Tal asked, raising his eyebrow.

"I'll explain all of that as well. You come too Sen. I can't believe Brake didn't tell me you were part of this."

Graham tried to eat slowly to avoid looking like Ming Li. She was shoveling potatoes in her mouth as fast as she could. Sen picked at his food and looked at them all suspiciously. Tal was eating steadily and watching his mom. It was funny to think Valeena was staying here in this modest little house, when she could be in a mansion in Akkron. The house was neat, but not showy.

"I guess you want to know why I'm here," Valeena said from the head of the table. "Shall I start at the beginning, so you will understand everything?"

"Yep. I'm listening," Tal said, leaning back on his chair.

"When I was young, I met your father. He was handsome and rich and I thought that was all I needed. It's embarrassing to admit it to you now, but those were the only reasons I married him."

Tal shrugged. "That isn't anything I didn't know."

"A lot of money was spent those first few years, and I thought I was living the best life. After you were born, something inside of me changed. I worried a lot more, and I wanted to protect you from whatever it was your father wanted you to become. The more I tried to intervene in your life, the more Briggs and I fought. He wanted you to be exactly like him," she said, grimacing.

"Traveling became more appealing to me. That's not the best way to handle things, but it seemed the best way at the time. The more I was away, the easier it was to leave. I can't stand being around Briggs anymore. I know that isn't fair to you, Tal." She placed a hand over Tal's.

"The traveling began to feel hollow, and I knew Brake was probably working with people to hunt The Dark Cloud. I know Brake and I knew he wasn't sitting in his house, ignoring the world. He wasn't hard to find. I told him I wanted to help. I figured out a way to part the clouds here, but it's very temporary. It has to be done every day, or there isn't any sun."

"So you figured you would never come back?" Tal asked.

"I was still going to come back, like I always do."

"You don't have to justify yourself to me. I'm used to having parents that don't care."

"I do care," she said, looking at her plate. "I've been a terrible mom. I worried that you were going to turn into your father."

"What? Why? I'm not a thing like him."

Ming Li rolled her eyes. "It's probably because you were walking around like, *hey I'm hot, and rich, and all the girls like me.*"

"I wasn't like that," Tal said, smiling and shaking his head. "And that was a terrible imitation of me."

Valeena looked at him with wide, hopeful eyes. "When you said you went into the cave, it made me realize you weren't going to turn out like Briggs after all. I'm sorry I wasn't around more to guide you or help you. I'll do better."

"I know you couldn't have taken me with you. Dad wouldn't have let you."

"What did he think about the cave opening for you?"

"He was ecstatic. He thinks I'm his key to becoming king of the world."

"I'm sure he does."

"We all went into hiding so people can't force us to open the cave."

"That's smart."

"And boring," Ming Li added. "That's why we came here. We're ready to help."

"Do you think Dad is part of The Dark Cloud?"

Valeena laughed bitterly. "No. He isn't disciplined enough to work with a group like them. They make plans and carry them out slowly and carefully. Briggs charges forward like a dragon, without a complete plan. He wants everyone to see him and admire him. The Dark Cloud doesn't advertise who they are."

"Dad told me he wants to send all non-magical people to Boztoll and make them farm for everyone else. He acted like they would have a choice, but I don't think he meant it."

"He said my mom could stay in Akkron," Ming Li said, shaking her head, "But that's because he likes her food."

"I'm not surprised," Valeena said, tilting her head. "Mali's Bakery is his favorite shop. I think he has some of the same ideas as The Dark Cloud, but I don't think he would join them unless he was the leader, and that will not happen."

"Maybe I can help you with the clouds," Ming Li said, setting her fork on the table. "I can make a pretty decent wind. If I could somehow get up high, I could try blowing the clouds away."

"It's something to think about," Valeena said. "We know there are Dark Cloud members living here, keeping the clouds around. We found one and captured him."

"Where is he?" Graham asked, excitement creeping into his stomach. "Maybe we could question him."

"This was a few months ago. Brake and Hamble came and took him away."

"Why didn't Brake tell us?" Ming Li grumbled. "I thought we weren't doing the whole 'secrets' thing."

"He probably didn't want us running off before he thought we were ready," Graham said.

*"I'm stuck in the cave!"* Wren's voice boomed inside Graham's head.

"Ahh!" He grabbed his head. Tal, Ming Li, and Sendo all jumped to their feet.

"What's wrong?" Valeena asked.

"Wren just screamed in my head."

"Is she okay?" Tal asked, gripping the top of a chair.

"Hold on," Graham told them. *"What do you mean, stuck?"* he whispered, picturing Wren at Meegore. *"And whisper please."*

*"The slide ended. I never made it to the bottom. I slammed into a wall."*

*"Can you open a portal?"*

*"No! I already tried. I can't do anything. I can't even make light!"*

*"Don't panic. I'm coming."*

"Wren's trapped inside the slide in the cave," Graham told them. "She can't make a portal. All she could do was call out to me." He stood up and started to fall. He caught himself on the table and stood with his head down until it stopped spinning. "I talked to her for too long."

"Great. It took all night before we all got over our headaches last time it happened." Ming Li said.

Valeena tapped her lip. "Wren is the fourth person who went into the cave? Drew's daughter?"

"Yes," Tal said, frowning. "Why did she contact Graham?"

"She might have tried you two," Graham said, pulling a few different bottles out of his cape pocket, "But I scrubbed all the rednax venom off of me when I was in the bathroom. She probably couldn't reach you." He found the one he was looking for and took the lid off. He took a swallow and tried not to gag.

"What's that?" Ming Li asked, wrinkling her nose. "It looks delicious."

"It's a medication I've been experimenting on and it's working already. I definitely need to work on the taste. It tastes like Brake made it."

"I ate one of Brake's cakes once," Valeena shuddered. "I've never made that mistake again."

"What's the fastest way to get to Meegore from here?" Graham asked. "She's scared and in the dark."

"I can take you there," Sendo said, looking uncertain. "It will only take a second."

"Can you take all of us?"

"I think so."

"Be careful," Valeena said.

Sen took a deep breath. "Everyone, hold on to my cape." Graham, Tal, and Ming Li all grabbed onto a piece. Everything spun. Graham thought he might be sick. It spun faster and faster, and when he could finally open his eyes, they were on the Island of Meegore.

Wren was trying not to panic. Everything was completely black. She couldn't even see her hand in front of her face. She could feel the darkness wrapping itself around her. Why had she come into the cave alone? When she hadn't been able to sleep outside on the cold, hard ground, she'd gone for a walk. When she'd passed near the cave, it had opened. She hadn't even had to put her hand on the crystal.

She'd looked around to see if anyone was near, and something crazy inside of her convinced her she could be brave and go into the cave by herself. Thankfully, she had practiced making larger orbs of light since the last time she was here, and so going down the stairs seemed a lot less scary than the first time. Once she'd arrived at the slide, she'd regretted her decision. She turned around to go back, but the stairs were gone. All that was there was a hard cave wall.

She rubbed her head. It felt like someone had used it as a drum. She had talked too much when she called Graham for help. When it became clear, the stairs weren't going to reappear, she'd gone down the slide. It was as scary as she remembered. Part way down, she'd remembered the river she was going to have to cross and panicked. Last time, Tal had taken her across on his back. She was also going to have to jump over the large pit, and no one was there to help her.

Without warning, she had plowed into a wall. The slide ended somewhere in the middle with no way to exit. Climbing back up had been a waste of time, but she'd tried. All that had come from that was scraped up hands and knees. It was terrifying in the dark. She tried to open a portal, and nothing happened. It wouldn't have been as bad if she could make light.

She's been worried she wouldn't be able to send a message to her friends, so when Graham answered her she almost cried with relief. Now she was going to have to sit here and wait. It could take forever if they had to fly from Boztoll. Wren thought of sending a message to Austra, but it would be pointless as Austra wouldn't be able to open the cave.

Something crawled across her hand, and she screamed, shaking it wildly. Why had she done this? Why did she think she had to prove to herself that she could be brave? She rolled back over onto her hands and knees and tried to climb up again. Her boots slipped, and she hit the wall. She sighed and took off her boots and stockings.

Now she would have better traction. She started back up. It was going a lot better this time. It wasn't good for her toes, but she was doing it. One step at a time. This wasn't so bad. Even if the stairs didn't reappear, it would be better to be stuck somewhere where she could make light and sit on flat ground until her friends found her. Just when she thought she had this, her foot slipped, and she slid back and slammed into the wall. She was going to be a walking bruise by the end of this.

"I give up," she muttered, pulling herself against the hard wall. She wrapped her arms around her legs and buried her face in her knees.

"How did you do that?" Austra demanded, walking to the place Graham, Tal, Ming Li, and Sendo stood. "It was like you appeared out of nowhere."

"We need to go to the cave," Graham said, not wanting to waste time. "Wren's stuck inside. Which way is it? I'm turned around."

"We woke up this morning and Wren was gone, and one side of the crystal was glowing," Austra explained, leading them through the trees. "It was an exciting thing to see, but she shouldn't have gone in alone."

"One of us will go in and get her," Graham said as they came into view of the cave.

"I wonder what happened," Tal said, looking at the purple glow coming off the large crystal. "It doesn't seem like you could get stuck in the slide. It was wide."

Sen gazed at the crystal, examining it. "Wow, that's really purple."

Hamble and Haltina came rushing towards them. Ten giants were pacing around the cave.

"We are worried," Haltina told them. "Wren must have gone into the cave."

"She sent me a message. She's stuck inside," Graham explained.

"I think we all need to go back in, not just one of us," Ming Li said. "Safety in numbers."

"Let's be smart this time," Graham said. "Is there anything we can use to sit on when we go down the slide?"

"Sit on?" Haltina asked, tilting her ancient head.

"Yeah," Ming Li said, her eyes sparkling. "Last time, some of us, who will remain nameless, ripped holes in their britches."

"These pants are a lot thicker," Tal argued, pointing to his jeans. "We shouldn't leave Wren there any longer than we have to."

"There was something in one legend of Meegore about getting stuck," Haltina said, scrunching her forehead. "It isn't something I focused on, so I can't remember much. I think it said that the cave would only let you go as far as you were prepared to go."

"That doesn't make sense," Ming Li said. "None of us were prepared at all last time and we made it through fine."

"I'm sorry I can't be of more help."

Sendo walked around the crystal. "I've heard about this my whole life, but I never thought I would get to see it."

"Your whole life?" Ming Li asked. "I know you're from Earth, so I doubt it's been your whole life."

Sen's head shot up and his eyes narrowed. "What do you know about Earth?"

"I'm from Earth. So is Graham."

"How did you get here?"

"We don't have time for that," Tal cut in. "The longer we wait, the worse it could be."

"Do we all need to touch the crystal again?" Graham wondered out loud.

"You shouldn't have to touch it at all anymore," Austra said, tapping her chin. "If you walk towards it, it should open, and your side of the crystal will glow."

"That's right," Haltina agreed.

"It looks so smooth," Sendo said, circling it again. "It has fingerprints all over one side."

Haltina nodded. "Yes. Just because the cave opened doesn't mean it can't be touched anymore. More people than ever have been coming to see it. Everyone only cares about touching the side that hasn't lit up, hoping to be the fifth. Once a side glows for someone, that is their side, so it's pointless for anyone else to touch them."

"And they touch, touch, touch, so we have to wash it," Tnarg said, coming up with a rag to wipe it clean. "I can't keep up."

"There isn't anyone around now," Ming Li said, looking through the trees.

"We had to change things, so now people need tickets. After it opened for you, it became very busy. We need some days to clean and have peace."

"Can I help wash it?" Sendo asked eagerly. "I love crystals. I have a collection back home."

"Of course," Tnarg said, handing him the rag. "Don't go from side to side. Only up and down."

"Does anyone have a rope?" Graham asked, getting back to the matter at hand. "If Wren can't get up the slide, we might need to pull her up."

Tnarg bowed. "I can go grab one. It will only take a moment."

"If we sit on our cloaks, that might help with the slide," Tal said. "Last time, mine was flying behind me."

"Whoa look," Sendo said, backing away from the crystal. "Now another side is glowing. Whose side is this?"

They all stared at the glowing purple light.

Haltina smiled at Sendo. "It appears it is your side."

# CHAPTER 5

I t was so cold. Wren pulled her cloak tightly around herself. How long had she been down here? It felt like forever. She wondered how long it would take her friends to get to her. Probably hours. She was getting a cramp in her leg. It was impossible to get comfortable sitting at this angle between the wall and the slide.

"Wren, are you there?" Ming Li's voice called from above.

Her eyes filled with tears of relief. "Yes!" she called. How had they gotten here so fast? Alicorns were fast, but not this fast.

"Did you go on the left or right side?" Tal's voice echoed down.

At least she thought that was what he said. She thought for a moment. "Right!"

"We're lowering a rope!"

Wren waited for what seemed like an eternity.

"Has it reached you?" Graham called.

She felt around. Nothing. She tried reaching up the slide and still couldn't feel anything. "No!" She was panicking again. If they couldn't get a rope long enough, what would she do? She could hear them talking above her, but she couldn't understand what they were saying.

"Stay to the right!" Graham yelled.

Did he mean his right when he was staring down, or her right going up? She moved to the side she would call right and waited. She could hear something, but she wasn't sure what. There were some clunks and scraping sounds. Someone must be coming.

"Wren?" Graham's voice called. He sounded close.

"Right here," she said, trying not to sound as scared as she was.

"The rope ran out. I can't tell how far you are. I'm going to hang down as far as I can. See if you can feel my foot."

Wren stretched up as far as she could go and reached out and moved her hand around. "I can't reach you."

"Whoa!" Graham yelled as he smashed into Wren, pushing her back into the wall. "Sorry," he said, moving off of her. "Are you okay?"

"I'm fine," Wren lied. Graham was not light. She wouldn't be surprised if she had a huge goose egg where her head hit the wall. She probably added a few new scratches as well.

Tal's voice echoed through the tunnel. "What happened?"

"It's fine!" Graham yelled. "Give us a minute!"

A light appeared in Graham's hand, and he held it up to inspect her. She shielded her eyes with one hand. It had been so dark for so long.

"Your head's bleeding a little bit," he said. "Does it hurt?"

"No worse than everything else. I'm pretty scraped up from all the times I tried to climb up. It makes little sense. The chute was so smooth, and now it's bumpy and sharp."

"Can you hold the light?"

"I'll try. I don't know why I can't make my own."

Graham handed her the light, and it didn't go out. He reached into his cape and pulled out a small bottle. He moved awkwardly around until he was lying at an angle on the slide. She followed his lead so she wouldn't be squatting. He untwisted the lid and turned, holding the bottle up to her mouth.

"Take a swallow."

She raised one eyebrow and frowned. "What is it?"

Half of his mouth turned up in what appeared to be an apologetic smile. "It's medicine. It tastes horrible, but it works pretty fast."

He put it close to her mouth, and she sighed, opening it. This felt ridiculous. He poured some of the worst medicine she had ever tasted into her mouth. She tried not to gag as the disgusting concoction made its way down her throat.

"That was awful," she shuddered.

"Yeah, I know. I need to work on that."

"My head is already better. Thanks. Now what do we do?"

"The rope isn't that far up. You were probably barely missing me. Let me rest my hands for a minute and then we can start up."

It was even more uncomfortable now, sharing the small space with Graham. It also felt weird, lying in this position, with nowhere convenient to put her feet. She hoped Graham's hands weren't too burned from holding onto the rope.

"Sorry you had to come for me."

"It's alright. That's what friends do." He sat up and ran his hand over the wall. "This is so weird. This is the same slide we slid down last time. A wall can't grow."

"Maybe there was some sort of rock slide?"

"I don't think so. It feels pretty solid."

"Did you get to the slides without any trouble?" she asked.

"Yeah. Same as last time." He shifted.

"I tried to go back when I came to the slide, but the stairs were gone."

"Gone?"

"Yes. A wall like this was blocking them."

"That's strange. It all seems normal up there. No rocks or anything."

"You don't believe me?"

"No, I totally do. This cave is magic. Who knows what can happen here?"

"Why would it be trying to trap me?"

"I don't know. Haltina said something about the cave reacting weirdly if someone wasn't prepared. I don't know, I wasn't paying a lot of attention," he admitted sheepishly.

"We should start back up. I can't stay in this position much longer. I'm sorry you had to be in here for so long."

"What should we do?"

Graham ran his hand over his hair and let out a slow breath. "It would be best if I try to climb to the rope. When I get it, do you think you can climb a little way?"

"Yes. Whenever I tried, I could get up a few feet before I slid."

"Okay, let's do this. You might want to put your boots back on."

Wren had forgotten she was barefoot. She put the light on her lap and quickly pulled her stockings and boots on as Graham climbed.

"Should I let the light go out?"

"No, try holding it up. I can see the rope up there. Once I get it, you'll have to let it go out. You have control of it now, so I bet you could make a new one if you wanted."

Wren held out her free hand, and sure enough, she made another light appear. Relief flooded through her. She'd been worried she might have lost her magic like her Uncle Karlof had. It wasn't common, but when she couldn't do magic, she'd been scared.

"Okay, I have the rope," Graham called down. "Climb up and grab my foot."

"I don't think this is going to work." Wren put the light in her pocket and kept focusing on it. It wasn't giving out a lot of light, but it was better than nothing. She climbed. "What am I going to do when I get to your foot?"

"I'll put my hand down and grab you. I anchored my foot to the floor, so grab my ankle. It might be better if we tie you to the rope and have Tal and Sendo pull you up, and I could come after."

"Who's Sendo?" Wren asked, willing her arms to stop shaking.

"Long story. Let's just say it's a good thing you didn't need us yesterday."

Wren could still make out shadows, and she could see Graham's foot. She reached out and grabbed his boot.

"Okay, now try to grab my left hand."

His hand was still too high, so she grabbed a handful of his pants and tried to pull herself up.

"Ahhhhh, grab my leg! If you pull my pants off, we'll be right back where we started."

"Sorry," she said, grabbing onto his leg with her other hand. She was glad it was dark enough he couldn't see the red creeping up her face. He reached his hand down and she grabbed it. He pulled her up to the rope.

"Great, now what?" she asked as she grasped the rope and tried to imitate the squatting position he was in. They were both holding on too close together, and it was way too wobbly for Wren's liking.

"Hold the rope and try to walk up. It's like scaling a mountain, but you'll have to stay squatting or you'll hit your head. I'll call up to Tal and they can pull. I'm going to slide back and wait."

"Okay, I'm ready."

"Tal! Pull!" Graham called.

Wren screamed as the rope pulled forward and dropped. They both went tumbling back. The slide was smooth and slippery again, and she didn't hit the wall. The light had gone out, and she was shooting down the slide backwards in total darkness. Wren tried to stop herself by dragging her hands on the sides, but that was painful. She closed her eyes and waited for it to end.

Graham shot out of the slide and landed on his knees. He did a quick somersault into the cavern so that Wren wouldn't fall on him. He heard her land next to him with a thud.

"Light," he said, and as he hoped, the space filled with light. Wren was lying on her back with her legs still on the slide. "Are you alright?" he asked, crawling to her.

She bit her lip. "Probably."

They could hear yelling from up above, but they couldn't make out what was being said.

*"Come down,"* he whispered to Tal. *"We made it all the way."*

Tal's voice filled his head. *"Give us a minute."*

"Wow, you're pretty scrapped up." Graham said, giving Wren a quick once over. Her face and arms were covered in scratches, and there were a few tears in her jeans. He was a little worried that she hadn't gotten up yet.

"Yeah," she said, glancing at him. "It feels nice to not be squished."

"You better hurry. The others will be down in a minute." He stood and helped pull her up and away from the opening.

"I don't know what I was thinking, coming here alone. How did the cave make walls appear?"

"Who knows? It makes me a bit worried about coming back too many times. If things can change, it doesn't seem safe." He pulled a bottle out of his pocket and poured a special lotion onto his hand. "Here, let me put some of this on your face. It will help the cuts." From the last few months of study, Graham had learned that any medicine he made would work better because of the magic he had obtained the last time he came into the cave. If he used medicine and was the one to touch someone, it also worked better. If he tried to heal someone with no medicine, it made him sick.

She cringed as he lifted his lotioned hands towards her. "I hope it doesn't smell as bad as that last medicine tasted."

"It doesn't have an odor," he assured her, rubbing it gently onto her face. It was awesome to watch the scrapes immediately start to fade. He rubbed some into her arms.

"It's amazing how fast your medicines work."

"Yeah, it's pretty awesome. I wish I could get my hands on some more medical books." He rubbed another spot on her cheek that he'd missed and then he took her face in his hands. The scapes were barely noticeable.

She tilted her head and stared at him with questions in her eyes. "What is it?"

Graham didn't know what to say. He didn't even know what he was doing. He swallowed hard. "I need to look at your eyes to make sure you don't have a concussion or anything."

"Oh. How can you tell?"

"When someone gets a concussion, their pupils are usually bigger than normal." Graham felt like an idiot gazing into her eyes, but he couldn't seem to let go or move. All he wanted to do was kiss her. He blamed Ming Li for all the kissing talk. He couldn't do it, though. They had all become good friends since they had been stuck together for so long. If he kissed her, it might ruin everything.

"Do they seem bigger?" Wren asked, raising an eyebrow. "I hit my head a few times."

"They look fine," he said, still cupping her face. Was he leaning? He must be, because Wren's eyes widened. Why did he feel like he was watching someone else? Had he lost all his senses? He needed to pull away now, but he seemed to be getting closer. Wren rested her hands on his wrists. He was sure she was going to push him away, but she leaned towards him. Only a couple of inches separated them, and he could feel her breath on his lips.

"Whoa!" Tal's voice called as he flew off the slide. Graham and Wren broke apart and Graham let out a breath. How had they not heard him coming? Tal landed on his feet and hurried out of the way as Ming Li fell to the ground, closely followed by Sendo.

"We're getting better at not landing on each other," Ming Li said, dusting off the back of her pants. They must have forgotten Tal's idea of sitting on their cloaks. "Are you okay, Wren? You seem a bit pale."

"I'm fine," she blurted out. Graham didn't look at her. He was way too embarrassed. He put his hands on his hips and walked in a circle. It was what he usually did when he missed a shot in a basketball game.

"Graham doesn't look so great himself," Tal muttered, clenching his jaw.

"He healed some of my scrapes," Wren said. "You know, that makes him a little queasy."

"It didn't seem like he was healing anything when I got here."

Graham noticed the lack of Tal's usual humor in his voice.

"I was making sure she didn't have a concussion."

"I'm sure you were."

"She's fine. Her eyes aren't dilated at all."

Tal glared at him. "Good to hear."

"What's wrong with you?" Ming Li asked, lightly shoving Tal. "You look annoyed."

"I'm not annoyed," he said, crossing his arms.

"Then why are you grinding your teeth and why do you look like you wanna punch Graham? He's the one who should want to punch you for dropping the rope."

"*I* dropped the rope?" Tal sputtered, throwing his arms into the air. "There is no way you can blame that on me."

"I would blame it on both of you," Sendo said, shrugging. "If you hadn't teased Li, she wouldn't have shoved you into me, and we wouldn't have lost the rope."

"Who are you?" Wren asked.

"This is Sendo," Ming Li beamed. "He captured us at Boztoll because he thought we were part of The Dark Cloud. He and his dad and brothers knocked us out and tied us up in this smelly old cabin. They kept acting all threatening and stuff. It turns out they're part of The

Silver Eclipse in Boztoll, which is led by none other than Tal's mom! Also, I'm pretty sure he's from Earth."

Wren bit her lip. "Oh. Nice to meet you. I'm Wren."

Sendo shook her hand. "Li isn't right. I'm not from Earth. I'm from Boztoll."

"Then why can you speak Spanish?" Ming Li asked.

He shrugged. "I don't, really. I only know a few words. My dad is from Earth, from Mexico. My mom is from Akkron. Me and all my brothers are from Boztoll. My parents live there since my dad can't do magic."

"Why is Sendo allowed to call you Li?" Tal asked, pushing his bangs out of his eyes. "You bite my head off whenever I say it."

Ming Li smiled and winked at Tal. "It sounds better when he says it."

"You've been acting weird ever since he kissed you."

Graham chanced a quick glance at Wren. She peeked at him quickly and away. "He kissed you?" she asked.

"Yes, and I've totally changed my mind about germs. He didn't do it because he likes me or anything. He only did it because Graham and Tal wouldn't, and it was my last request."

Graham tried not to smile. He thought back to a conversation he and Tal had eavesdropped on. Ming Li and Wren had been talking about kissing, and Ming Li said she was worried about getting someone's germs in her mouth. It seemed Sendo had helped her overcome that fear. Sendo was hard to read, he was standing there looking stoic.

"Graham and Tal wouldn't kiss you ...?" Wren cocked her head and focused on her friend.

"Are we going to stand around talking all day or get out of here?" Tal asked. "There are people waiting up top to find out if you're okay."

"So, you brought him so he can get magic?" Wren asked.

"I already have magic," Sendo said, "And you can all call me Sen."

"He's our number five," Ming Li grinned. "The crystal lit up for him."

"Oh, that's a good sign, isn't it?" Wren asked, glancing at Graham and then quickly to Tal.

Graham rubbed his curly, dirt covered hair. "Probably. Last time we were here, Haltina said if we all came out of the cave together, we would be bound together, or something like that."

"Why are there two hallways?" Ming Li asked, pointing. "I swear there was only one last time."

"There was," Graham said, analyzing the pathways. "So, which one do we pick?"

"It probably doesn't matter," Tal muttered. "I'm guessing it all ends up in the same place like before."

"What's with the yellow dust?" Sen asked.

"We never figured that out," Graham said, kicking at the dust.

"I could make a portal out," Wren offered.

"It probably won't work," Graham said, walking towards one opening. "There are probably precautions against stuff like that. Whoever was in charge of this cave wouldn't have wanted people popping in and out. Besides, we're already here. We should at least take advantage of it, don't you think?"

"We should," Tal said. "Let's go that way."

"Come on Sen," Ming Li said, pulling the boy by his arm. "Just wait until you see the room full of magic."

Sen and Ming Li were going fast, and Wren was right behind them. Tal was strolling with his hands in his pants pockets. Graham walked beside him, feeling awkward.

"What's going on?" he finally asked. Ignoring it wasn't going to make it go away.

"What are you talking about?"

"You were fine when I slid down the slide, and now you're acting all irritated."

They walked in silence for a minute. Graham figured he wouldn't answer.

"It's not what you think," he finally said. "It's just ... I don't know. We have this whole destiny thing with the five of us. We have to stick together. Work together. Now it's obvious Li is crushing on the new guy, and you and Wren are down here kissing–"

"We weren't kissing," Graham cut in.

"Well, you probably would have been if I hadn't come when I did. I've never seen two people look so guilty in my life. And don't give me the whole concussion story. You were way too close to see whether her eyes were dilated or not."

Graham let out a long breath. "I don't know what came over me."

"I do," Tal said quietly. "Wren is pretty great. I'm not blaming you or anything. You can kiss whoever you want. What if these relationships form and something goes wrong? We still need to save the world. We can't afford someone getting bitter and running off."

Wren gazed back at them, and Graham sighed. "I know. You're right."

"I'm not only saying this because I'm jealous. I mean, I'm totally jealous, but if that was all, I would ignore it." His lopsided smile was back.

"Why would you be jealous? Aren't you the one with the arranged marriage?"

"Ugh," Tal laughed, shaking his hand. "I hoped you forgot about that."

"Never," Graham said, raising an eyebrow. "So, who is she?"

"I'm not supposed to talk about it until it's announced some day. That's why it's hard to date in Akkron. You never know who might have an arrangement. I don't know much about her. My parents arranged it when I was, like, seven."

"Can't you refuse?"

"Sure, but I can't ever marry anyone else, and they can arrest my parents for breaking a contract." He paused and then chuckled. "Hey! I might have figured out a solution for my dad."

"So there's no way out?"

"There is one. If the girl refuses, she can have the contract broken, if her parents agree."

"That's weird and unfair."

"Yeah, and it never happens."

"How many people have arranged marriages?"

"I don't know. Probably about one-fourth."

"One-fourth? That's huge!" Graham couldn't imagine having his parents arranging his marriage. That wasn't surprising, since he couldn't imagine having parents. He wondered if Wren had one. Probably not, or Ming Li would have mentioned it by now.

"Yeah, it isn't as popular as it used to be. Now it's usually upper class snobs that want to control their kids."

"Is there a way to change the law?"

"I don't know. Wren's parents had an arranged marriage."

"Really?"

"Yeah. I've heard Wren talk about the way her dad won't talk about her mom, and it's probably because they didn't know each other that well. I mean, they worked together and stuff, but from what my dad said, they weren't in love or anything."

"Does Wren know?"

"I doubt it."

Graham pointed ahead. "There's the wall we have to climb." Wren, Ming Li, and Sen were all standing at the bottom, staring up.

"That is so high," Wren said, biting her lip.

"Do you want to race to the top?" Sen asked Ming Li.

"Do you want to lose?" she smiled. "Ready, set, go!" Sen and Ming Li both started climbing. Ming Li was fast, but she had nothing on Sen. He scurried up so fast, Graham worried he might fall.

"You're pretty fast," he said to Ming Li as she joined him on top.

"It's about time someone beat you," Tal said with a grin.

"Maybe I let him win."

"Nope. You don't let people win. And I've raced you. That's as fast as you can go."

"Well, at least you've never beat me," Ming Li said, sticking out her tongue.

Graham glanced at Wren. If she frowned any harder, her face might crack.

"Are you alright Wren?"

"I'm never going to make it up. This is way worse than Coach Zalliah's wall. There aren't many hand holds, and it's too tall for you guys to pull me up."

"It's only about twenty feet," Tal said. "Graham can climb to the top first and help you when you get to the top and I'll go behind you in case you fall."

Graham nodded and started climbing.

Wren sighed. "If I fall and you're behind me, I'll knock you down as well."

"I'll go to the side of you, just a little lower. If you feel like you're going to fall, tell me."

Graham wondered if Tal wanted a chance to be the one helping her. He knew it shouldn't matter. He kept climbing. It was harder than Zalliah's wall, but it wasn't terrible. He reached the top and stood by Ming Li and Sen. Wren was coming steadily and doing a lot better than she would have three months ago. All of her training was paying off.

"You're doing great," Tal told her.

When she came close to the top, Graham bent down on his hands and knees and reached for her. She could probably make it without him, but why risk it? She grabbed his hand, and he helped her over the top.

"Thanks," she said, brushing the dirt off of her cloak. Tal was right behind her. "There's a small river up ahead," Ming Li said, glancing sideways at Sen. "You wanna race again?" Sen smiled and took off running. Ming Li was only a step behind.

Wren let out a slow breath. "Is everything going to turn into a competition?"

"Maybe," Graham said, watching Tal.

"Only between those two," Tal added. Graham hoped he was right.

# CHAPTER 6

Wren peered into the murky river and sighed inwardly. This was even more nerve-racking than the last time they were here, because now they knew there was something in the water. Tal had felt it rub against his leg. It hadn't affected Wren, because she'd ridden across on his back, avoiding the freezing water and whatever creature lived inside. She was probably going to cross on her own this time. It was one of her goals to be stronger and less dependent on everyone else.

"Last time we were here, I just plunged in," Ming Li said, rubbing the goosebumps on her arms. She peered wearily down into the opaque river. "Talon said there's a snake or something in there. I don't really do snakes."

"I don't sense anything," Tal assured her, bending close to the water. "It's possible I freaked myself out last time."

"I could try levitating everyone over," Graham offered.

Sen shook his head. "It won't work."

"Why not? We can make light."

"Haven't you studied this cave? There's a limited amount of magic that will work here. The entire process shows your dedication. If people could fly through, it would defeat the purpose."

"We should have talked to Haltina before we came back in. I thought we would be more prepared the second time around."

"Sorry," Wren said, staring at her feet. It was her fault they were in here unprepared again.

Tal shrugged. "It's alright. We didn't do that well in Boztoll, and we had a plan."

"Let's go," Graham said, adjusting his cloak. "Waiting won't make the water any warmer."

Tal turned to Wren. "Do you want to get on my back again?"

She cringed and shrugged. She really did. The thought of wading through freezing cold water that might have something living in it was horrible. She also didn't want to be the only one who didn't do it on her own.

"There's no point in everyone getting wet," Graham said. "I tried to convince Ming Li last time, but she was being stubborn."

"And I regretted it," Ming Li said, shaking her head. "It was soooo cold. I'm totally willing to get a piggy back this time."

"Okay, I'll take Wren and you take Ming Li," Graham said to Tal.

Tal's eyes flashed with something Wren couldn't understand, and he clenched his jaw. Wren wanted to go with

Tal. Being with Graham made her insides twist up, but she didn't want to make everything awkward by refusing.

She felt pretty sure he would have kissed her back there if Tal hadn't arrived at that moment to interrupt destiny. Unless he truly had been examining her eyes. Just the thought of being close to Graham right now made her heart pound. She had been trying to avoid looking at him since it happened. The worst part was the empty feeling that seemed to have settled in her stomach. She worried it might stay there forever.

"Hello? Wren?" Tal was saying, waving a hand in front of her face.

"What? Sorry, I spaced out for a minute."

"They want you to tell them who *you* want to take you across the river," Ming Li said, rolling her eyes. "At least that's what they've been arguing about for the last few minutes."

They were arguing over who should take her? That made little sense. Why would they even care? Ming Li would be easier, because she was shorter and lighter.

"Li can come with me," Sen said, shaking his head at Graham and Tal. She smiled and jumped on his back, and they plunged into the river.

"So, who do you want to go with?" Tal demanded.

"It doesn't matter."

Tal and Graham scrutinized each other. Tal was clenching and unclenching his fists, and Graham was grinding his teeth. What was going on? Tal and Graham always got along. Something must have happened in Boztoll.

"I'll go myself," Wren said, raising an eyebrow. "I don't know what's gotten into you two." She walked up to the water, but before she could step in, Tal pulled her back.

"It's extremely cold. Go with Graham."

Graham nodded and squatted so she could climb on his back. Wren shook her head in defeat and climbed on, holding her cloak and Grahams up so they wouldn't get wet. She didn't want to find out what was in the water. So much for showing her independence. Tal was already trudging through the water. Graham stepped in and shivered. The water was up to his knees. Ming Li and Sen were already on the other side.

It was so weird being this close to Graham. Wren wondered if she should talk or something, but she had no clue what to say. She tried to keep her face away from him, even though something in her wanted to be closer. She scolded herself for even thinking about it.

"Don't fall off," Graham said, through chattering teeth. "It seems like you're trying to. Relax."

Wren willed herself to relax. She held on tighter and leaned her face over his shoulder. Her cheek touched Graham's, and she jolted back so hard she let go. The next thing she knew, she was upside down in the coldest water she could ever imagine. Graham let go of her legs and she flailed. She touched the ground with her knee and pushed herself up. The first breath of air stung her lungs, and she coughed as she tried to stand.

Graham helped her stand the rest of the way, and she noticed water running down his face.

"Did I pull you in?" she asked, horrified. Her whole body was numb.

"It's alright," he said, clenching his teeth to stop the chattering. "Let's get out of here."

He grabbed her hand, and half pulled her towards the others. She peered up at Tal, Ming Li, and Sen. They were all staring at them with their mouths gaping, and then Tal laughed. Ming Li shook her head and frowned. When they came to the end, Tal reached a hand out to both of them and helped them step out of the water.

"Take off your cloaks," he ordered, helping Wren peel hers off.

"Wow, it's cold," she said, as Tal wrapped his dry cloak around her.

Sen was rubbing his hands together as fast as he could and gazed intently at the two of them. "Stand by each other," he commanded. "You too," he said to Tal. He put his hands out and Wren moved closer to the heat coming from them.

"How are you doing that?" Graham asked.

"Practice. I'll try to show you sometime. You might want to turn in slow circles. You don't want to get burned."

Wren slowly turned, and within five minutes they were dry. That was definitely a skill worth practicing.

Graham inspected his dry clothes. "That was crazy."

"I've never seen anything like that before," Tal said, studying his dry clothes.

"Yeah, there are some things we don't want people to be aware of. It gives us advantages against The Dark Cloud."

"Well, we're almost there. Should we go?" Tal asked. "Perhaps Graham and Wren should stay behind and iron out some of their issues."

Wren's eyes narrowed. What was he talking about?

"What issues?" Ming Li asked. "If anyone is having issues, I would think it's Graham and Tal. You two are acting weird."

"Let's go," Wren muttered.

"Are you sure?" Tal asked. "You can catch up if you want to."

Wren glanced up at Graham and away.

"Not ready for that? Well, if you don't talk things out, you're going to keep having these weird moments."

"I don't know what you're talking about," Wren said, walking into the hallway of the cave. "Let's go. We don't want to keep Haltina waiting."

Ming Li caught up to her and linked their arms.

"What's Tal talking about?" she whispered. "And why does Graham look like you punched him in the stomach?"

Wren winced. Did he actually want to talk about it? She shifted uncomfortably. She was positive she wouldn't live through a conversation like that. What would he even say? And what did Tal know about it? Did Graham tell him? That was ridiculous. There wasn't even anything to say. All he did was look into her eyes, and lean in incredibly close.

"Whoa," she said, as she tripped over her own feet. She managed to stay up and not take Ming Li with her.

"Are you alright Wren?" Tal asked from behind them. "I'm sure you can still change your mind and have that conversation with Graham ... OWWW! Pinching? Really? What kind of wimp move is that?"

Flames climbed up Wren's neck and onto her face. She grabbed Ming Li and hurried them forward.

"Something is totally going on," Ming Li said as they entered the final cavern in the cave. The silver walls glistened and the colorful bubbles floated around in the air. In the middle of the room was a four foot stand with a thick old leather book on the top.

"Where did that come from?" Wren asked, hurrying over to the book.

"Wow," Sen said, walking around, inspecting the room.

"What does the book say?" Graham asked, walking up to Wren.

Her hands shook as she touched the brittle cover. *"The Silver Eclipse."*

"The Silver Eclipse?" Sen asked, glancing over Wren's shoulder. "Is it like instructions or something? If we can make a silver eclipse, that could save Boztoll."

Graham was excited. This might be what they were waiting for.

Wren opened the book and turned the pages. "It appears to be a recipe," she said, holding it up to Graham.

"I wish I had a way to take a picture," he muttered, trying to memorize the page. There were a lot of steps.

"Do you suppose we can take it with us?" Tal wondered.

"I don't think we should," Wren said. "We don't want to get in trouble."

"I have a pen," Sen said. "I'll write it on my arm."

Wren put the book on the stand so Sen could see it. He started writing.

"That's a thick book," Graham said. "Do we need the whole thing?"

Tal shrugged. "If we do, we can come back."

"I'm a little nervous about coming back after being stuck," Wren admitted. "And it might not be here next time. It wasn't here the first time. How did you all get to me so fast? I was sure I would be waiting a lot longer."

"Sen can teleport," Tal said.

"Wow, I thought that was lost magic."

Sen shrugged. "It is, but The Dark Cloud and some people in Boztoll rediscovered it. I can move around this world, but I never learned to go to other worlds."

"So The Dark Cloud members aren't disappearing, they're teleporting." Graham said, rubbing his chin. It would be nice if Sen could teach them all, so they wouldn't have to rely on Wren's portals for everything.

"I'm taking a blue bubble this time," Ming Li said, pulling a bubble out of the air. Last time they all received the magic they needed, or wanted. Graham wondered if the color of the bubbles mattered, or if the cave somehow took their needs into consideration. Great. Now he was thinking of the cave as a living being.

"Okay, I've got it all," Sen said, holding up his inked arm. "The rest of the book seems to be a history."

"Let's hurry," Ming Li said. "Thanks to Sen and his brothers, I'm tired."

"Sorry," Sen shrugged, unapologetically. "You guys need to work on your stealth."

Graham grabbed a white bubble, Sen a red, and Tal jumped and snatched a pink.

"I guess after this we'll know if the colors make a difference," Wren said as she caught an orange bubble.

"So, Haltina said if we all came out together, we'd be bound together," Tal said. "What does that mean?"

"I guess we'll find out," Graham said, walking up to the raised platform in the middle of the room.

"Should we all hold hands?" Ming Li asked.

"Probably," Tal said. "It was pretty crazy last time."

"What's going on?" Sen asked, stoically. Graham wondered if this guy ever showed emotion.

"We stand on that platform and we swirl away, back to the island," Graham explained. "It was kind of like when you teleported us to the island, except there were tons of colors swirling around us."

Ming Li grabbed Sen and Tal's hands, and Tal grabbed Wren's. Graham swallowed. He either had to take Sen's hand or Wren's. He sighed and took Wren's. She stiffened almost imperceptibly. Why did he have to make everything weird? Next time he would give her the lotion and let her put it on herself ... although it might have been worth the awkwardness if he had actually kissed her.

"Ready? Let's go," Ming Li said.

They all stepped up. Rainbow swirls surrounded them, and they were flying through the air. It wasn't as disorienting as the last time since they were expecting it. It wasn't as cold either, since they were dry. As soon as Graham's feet hit the ground, he swerved around to make sure no one was there to hit him in the head like last time. Thankfully, all he could see were trees and purple crystals.

Sen looked around. "That was fun."

"You don't seem to get impressed easily," Ming Li observed.

He shrugged. "My mom always says I seem bored. I'm not. I guess I don't show a lot of emotion."

"There they are!" someone called.

It wasn't long before Austra, Hamble, Haltina, and an assortment of other giants surrounded them. Everyone was asking questions. Tal seemed to enjoy answering, so Graham let him. His eyes met Wren's, and hers widened. This was going to be ridiculous. What if they worked together forever and he couldn't even talk to her anymore? He had to act normal. He smiled at her, and she looked confused.

"What are you doing, Graham?" Ming Li asked. Everyone turned and stared at him. "You look like you took a bite of Brake's food, and you're trying to pretend you don't want to barf."

"Are you well?" Haltina asked.

"Fine," Graham said. "I'm going to go check on something." He was such an idiot. This was getting worse and worse. He turned around and tripped over a fallen tree branch. "I'm fine!" He popped up and walked away. He walked until he came to the clearing with the crystal. All the sides were dull now. He sat on a flat crystal and sighed.

"Hey," Tal said, sitting on the ground next to him. "Sorry about the way I acted in the cave. Right after I told you we weren't competing, I started competing. I don't know what comes over me sometimes."

Graham sighed. "It's alright. I should have let you take Wren across the river. She wouldn't have freaked out and pulled us in."

"What happened anyway?"

"I don't know. She was trying to keep her distance, which doesn't work when you're on someone's back. She got too close and jerked back."

"I should have waited one more minute before I came down the slide and you would have had time to–whatever. Then you both wouldn't be acting all klutzy and weird."

"I don't know," Graham said, putting his face in his hands. "It might have made it worse, though."

"She's right through there," Tal pointed. "You can go try again."

Graham ignored his suggestion. "I'm blaming this all on me being tired. Spending the night on that dirt floor wasn't comfortable."

"I slept fine."

"Yeah, you were still out of your mind from the ailam powder. You had nothing to worry about."

"My life has been pretty strange since you came. It's feeling normal to be in peril one moment and fantasizing about kissing girls the next ... Of course, no one is *actually* kissing girls, except Sen."

"Hey, you had your chance," Graham smiled.

"Ming Li is one of my best friends and all, but kissing her would be even worse than you kissing Wren. I've put her in the category of a little sister that I like to tease."

"So, do you think we get a break after today? Last time things became exciting, I didn't feel like I was ever going to rest. Things got crazy."

"I doubt it."

Wren, Ming Li, and Sen raced into the clearing.

Wren looked excited. "Haltina told us the name of the person Brake was after on Earth! He owns some big company in a place called California. She's pretty sure he's part of The Dark Cloud even though he's still on Earth. I need to tell you all about the things the adults were doing when they were young. They told me about it last night."

Wren told them about Brake, Zera, Hamble, Drew, Tram, Austra, Fria, and Magnalee's work with Governor Jorin. Graham was surprised to hear about Fria's betrayal, and he couldn't believe Magnalee's manner of death had been kept from Wren all these years. It seemed like something everyone would talk about.

"So, why are you so excited about knowing the guy's name?" Tal wondered.

"Because we can capture him."

"The adults couldn't even capture him," Graham scoffed, shaking his head.

"Yes, but now nobody has bothered him in years. Not since Ming Li came here. He's probably let his guard down. He wouldn't be expecting us."

"If he's on Earth, he's not hurting anything here," Sen said. "Shouldn't we be focused on the people here?"

"He's probably still connected, though. I think people are communicating between here and Earth more than we know."

"Haltina said there has always been a link between here and there," Ming Li said. "That's why we cross with Earth so much."

"So, what do you want to do?" Graham asked them.

"My opinion? We should wait until tomorrow and then go to Earth," Wren said. "Austra said that Zera is at the meeting place, so we can go now and tell her. Then we can get a good night's sleep and we'll be ready tomorrow."

"Do you think she'll let us go?" Tal asked.

They all thought about that for a minute.

"Aren't we supposed to be the ones chosen to fix this mess?" Sen asked. "I'm still getting a feel for this whole chosen thing, but I feel like that means we don't need permission."

"They are part of our group though," Graham argued. "We can't violate trust."

Sen held up the arm that was covered in writing. "What about the silver eclipse?"

"There are a lot of weird things about that. When we get back, we need to copy it on paper. We're going to need all the help we can get to figure it out."

"I hope you five are not planning to do something stupid," Austra said, joining them.

"We're planning on going to the meeting place for a rest," Graham told her.

Her eyes widened. "Oh, well, that seems sensible of you."

"Did Haltina tell Wren everything she wanted to?" Tal asked. "It seems like that was too fast."

"She wanted you all to realize the connection between this world and Earth. She thinks it might be important."

"I'm surprised she had me come here just for that," Wren said. "I guess it might be useful, but she could have sent it all in a message."

"Perhaps she was planning on telling you more, but realized you were all a bit too impulsive."

"Impulsive, us?" Tal grinned. "Who would call us impulsive?"

Austra rolled her eyes. "I assume anyone who has met you."

Wren crawled into her cozy bed in the meetinghouse and pulled her down comforter over her legs. The rough-hewn wooden bed posts cast flickering shadows across the room, and she could hear the wind rustling outside. She glanced over at Ming Li, who was in her own bed, a funny look on her face that Wren couldn't place.

"What is it?" Wren finally asked.

The covers ruffled softly as Ming Li sat up in her bed. "I have been soooo patient, and you know that isn't my thing. Spill."

Wren sighed, feeling the weight of her friend's scrutiny. "What are you talking about?" she asked, waving her hand to turn out the light.

"You know what I'm talking about," Ming Li persisted. "What's going on with you and Graham ... and possibly Tal."

"Nothing."

"Please. We've been best friends for how many years? Something happened."

Wren let out a long breath. "I don't know what's going on with Tal and Graham. They usually get along really well."

"Tal was perfectly normal when we entered the cave." Ming Li said, tapping her chin thoughtfully. "Well ... as normal as Tal is. When we got to you, he was totally different. He was all irritated and stuff. Then you and Graham acted all weird. What was going on when he came down?"

"I plead the fifth."

"You can't plead the fifth. This is not The United States, and I'm the one who taught you that expression. If you refuse to talk, I'll come to my own conclusion."

"Ugh," Wren muttered, putting her pillow over her face. She wanted to scream.

"I could go ask Graham. Or Tal."

"Sometimes you're annoying. I had a bunch of cuts on my face, so Graham was rubbing some of his lotion on them. Not a big deal."

Wren could almost feel Ming Li raising her eyebrow. "I don't get why that would make Tal mad."

Wren's cheeks burned. "Well, he was also staring into my eyes to see if I had a concussion."

"Staring how?" Ming Li pressed.

"How many ways are there to stare?"

"Plenty."

"Well, he was extremely close to my ... uh, lips," she admitted, the heat in her cheeks becoming a small inferno. "He had his hands on my face. I don't know why Tal's mad, though."

"Graham was going to kiss you?" Ming Li squealed.

"Maybe ... uhhh! Do we have to talk about this?"

"Were you going to let him?"

Wren hesitated, her mind a whirlwind of conflicting emotions. "No. Yes. Probably."

"Okay. It all makes sense now."

"Does it?" Wren asked, staring up into the darkness.

"Of course. Graham was about to kiss you, but was interrupted by Tal. That's why you two have been acting weird, and Tal is acting weird because he's jealous."

Wren frowned. "Jealous? That doesn't make sense."

Ming Li laughed. "It makes total sense. Tal and Graham both have a crush on you."

"They do not," Wren insisted with a small giggle.

"Oh, come on Wren. It's been obvious since we were thrown together that they both like you. It kind of irritated me at first, but they are totally off my crush list now."

Wren was taken aback by Ming Li's candor. "Why is that?" she asked.

"Because ever since Sen kissed me, he is the only one on my list."

Wren's eyebrows shot up. "I need to hear that story."

Ming Li chuckled. "Later. Right now we're fixing you."

"I'm not broken."

"You and Graham can't be awkward around each other forever," Ming Li said firmly. "Me and Sen aren't awkward. Of course, he only kissed me as my last request. I don't get embarrassed that easily. You ... I think you live to be embarrassed."

Wren shifted. "Can we talk about anything else?"

"Not until we fix this. You're not going to feel less weird with Graham unless you kiss him."

Wren's eyes widened in mortification. "No way! That would make it way worse."

"Why?"

"It just will." Wren struggled to find the right words. "I want to pretend like it didn't happen."

"Do you like him? More than Tal?"

Wren frowned into the darkness. She did like Graham, but she liked Tal too. Tal didn't make her feel weird, though. "I'm pretty sure we are all too young to be worried about all of this, anyway."

"It's not like I'm asking you to marry one of them."

"I want to ignore it for now."

"So you're going to trip around and avoid contact whenever Graham's around?"

"No. I'm going to sleep and tomorrow it will all be back to normal," Wren replied with more certainty than she felt.

"You can keep telling yourself that."

"I'm glad I have your permission," Wren shot back.

"Perhaps I need to talk to Graham and Tal."

"No!" Wren said, shooting up. "Promise not to talk to them about this. Tal doesn't even like me. I'm not even convinced Graham does. Promise me you won't say any-thing."

"Okay, I promise. But that doesn't mean I won't be thinking about it."

Wren sighed, overwhelmed by the mess of feelings swirling around inside her head. Graham probably wasn't even trying to kiss her. Maybe he needed to get close to see her eyes. Graham couldn't like her, could he? Neither could Tal. They both smiled at Ming Li more than they did her. It had always bothered her, the way boys gravitated toward Ming Li and ignored her. It would be nice to have a boy like her for once, especially a boy she liked. Growing up was hard.

# Chapter 7

Everyone's sore muscles put the plan to go to Earth on hold. They spent the day reading and re-reading the instructions to create a silver eclipse and trying to figure out what it all meant. So far, they weren't having a lot of luck. To top it off, Zera wasn't around, so when they ran into problems, they didn't have anyone to ask. Wren felt like they'd wasted the whole day.

On the plus side, she'd acted completely normal around Graham. Perhaps she'd just needed sleep. Ming Li told her about their time in Boztoll, and Wren couldn't help feeling relieved she hadn't gone with them. Now that they had rested for a day, they were ready to take on Earth. In a rare show of responsibility, they decided to find Brake first and let him know what was happening.

"Brake's at Wren's house," Graham said, joining the other four in the kitchen for breakfast.

"My house? Why?" Wren asked, brushing muffin crumbs off her lap.

"I guess people take turns going there to dust and check on things while your dad's gone."

"Oh. That's nice of them," Wren said, letting out a relieved sigh. She'd wondered what was happening to her house, but had been trying not to worry about it. She should've asked someone and saved herself some stress.

Tal stood up. "So, let's go talk to Brake."

"I'm going to stay here and keep trying to work out what this means," Sen said, holding out the instructions to the silver eclipse.

"I'll stay and help," Ming Li volunteered.

"Did Li say she wants to search through books?" Tal gasped.

Ming Li glared at him and didn't respond.

"So the three of us?" Graham asked, glancing at Wren and Tal.

Tal nodded. "Looks like it."

Wren opened a shimmering silver portal to her house and held out her hands, allowing Graham and Tal to each take one. They stepped through the portal and into her front yard.

"I've missed this place," she said, pulling them towards the white stone house. "Of course I miss my dad more."

Wren pulled on the front door, and it opened easily.

"Isn't that your friend, Tal?" Graham asked. Wren turned and saw Solia in the distance. Not the person Wren wanted to see.

"She must be here for Wren," Tal said, bouncing into the house. "I'll check this floor for Brake and you look on the second floor, Graham."

As the boys hurried away, Wren shut the door behind them and waited politely for Solia. She sighed. What could Solia want here? She'd never liked Wren. She was probably here to see if anyone knew where Tal was. The closer she came, the harder it was for Wren to not go inside and shut the door.

Solia always appeared perfect. Her straight blond hair cascaded neatly to her waist, and her pristine blue tunic looked like it had never seen dirt. The only thing spoiling Solia's appearance was her hard blue eyes and the scowl on her face.

She stopped in front of Wren and pushed a strand of hair behind her ear. "Wren." She said her name like an accusation. "It's about time you were here. I was beginning to think you were hiding from me."

"Why would I hide from you?" Wren asked, confused. Of all the people she was currently hiding from, Solia was not on the list.

Disdain practically dripped from Solia. "I know what you're up to."

This was ridiculous. Wren didn't know how to respond to this girl. "Um ... I'm pretty sure you don't know what I'm up to."

"I know you have some pretty high and mighty ideas, now that you're one of the chosen. I'm here to let you know that you being chosen doesn't make any difference in some things. Sure, everyone is talking about you and the others, but in the end, you are still lacking." She looked Wren up and down.

"I have no idea what you're talking about," Wren said, hoping she wasn't visibly shaking. She hated confrontations, and this one didn't even make sense.

Solia's eyes narrowed. "I want you to know that no matter how chosen and special you believe you are, Tal will never fall for you."

A small giggle escaped Wren, and she shook her head in disbelief. "You came here to tell me I can't have Tal?"

Solia put her hands on her hips and clenched her teeth with an audible crack. "I know you think you're better than everyone else. I can't tell you how much I hate hearing everyone at school talk about your hair or your eyes. It makes me want to barf. Sure, Tal might have joined in the talk a time or two, but that doesn't mean he actually likes you."

Wren stared at the angry girl in amazement. Solia was accusing her of thinking she was better than everyone else? And what did she mean by *all the boys at school talk about her?* If that were true, some of them would have actually talked to her.

"I saw Tal go into the house. Go and get him for me," Solia demanded.

Wren stiffened. "I'm not your errand girl."

"Just go get him." She waved her hand dismissively. "I'm sure he's missed me."

"We're kind of busy," Wren said through gritted teeth. "Why don't you catch him another time?"

The door opened, and Tal came out. "Hey, Solia. How's it going?" he draped a strategically casual arm over Wren's shoulders.

"Where have you been?" Solia huffed. "You haven't checked in with anyone. Aren't you at all concerned about your friends and your father who have no idea where you have been?" she crossed her arms and glared at him.

"My friends know where I am," he said, shrugging. "Sorry I can't catch up, but we have things to do. Wren, can you come help me for a minute?"

Wren thought Solia might explode right there. Her eyebrows were knitted together and her face was red. "Sure."

"Thanks," he said, giving Wren a quick peck on the cheek. She tried to appear natural and not like she was about to die right there in front of them. Solia turned and stomped down the walkway. Wren and Tal hurried inside and shut the door.

Tal burst into laughter. "That was perfect."

"Perfect?" Wren asked, her fingers nervously twisting the hem of her shirt. Her face felt unnaturally warm. "What in the world were you doing?"

Tal's eyes crinkled with amusement. "Solia has been a pain in my neck for so long. I'm hoping she'll get the hint and stay away."

"You didn't have to kiss me," Wren protested.

"Oh, come on! Wasn't it worth it to see the look on her face? I'm putting it in my top five memories. It's right behind the time you fell on Jaaz's head and he pretended to eat you."

The corners of Wren's mouth quirked up. "Okay, it might have been nice to see her lose for once, but still."

"I'm sick of people who think they are better than everyone else. Solia was the worst. It's been so nice hanging out with our new group. I used to get migraines all the time. I haven't gotten a single one since I stopped hanging out with those people."

"It gave you migraines? That's strange."

"I think it was from trying so hard to fit in where I didn't belong. It takes a lot of energy to be someone you aren't. I'm so glad I forced you all to be my friends."

"We weren't forced," Wren said, grinning at him. "We were just pushed a little."

Tal laughed. "I guess my dad gets some credit there. I'm glad you got the push."

"So am I. I truly am sorry for the way I judged you. You've been a great friend these last few months. It makes me wonder how many other people could be beneficial to my life if I just gave them the chance."

"That's probably true," he said thoughtfully. "It's easy to get in our comfort zones and not branch out. If I would have tried earlier, I could have made different friends and been a lot better off."

"Well, we all have each other now."

Tal winked. "And now we know why Solia has always hated you."

"We do?"

"Yeah. She's jealous. I always knew she didn't like you, so I tried to talk you up to her. It never seemed to help. I guess it was only making her hate you more. Solia only likes the world to revolve around her."

"I don't know why she would be jealous of me. She's the pretty one, with all the friends and stuff."

"Seriously?" Tal asked, shaking his head. "Solia is okay, but you're beautiful. You heard what she said. All the boys at school are always talking about you."

Wren's stomach was doing flips. That couldn't be true, could it? "If that was true, then boys would actually talk to me. They all ignore me and talk to Ming Li."

Tal stopped smiling and regarded her. "That's really how you see things? Ming Li is easy to talk to. She likes sports, and she talks to everyone. People like her, but people like you too. You just ... I don't know how to describe it. You're a little scary?"

Wren could feel her lip trying to tremble. "Scary?"

"I'm not messing with you, Wren. You're pretty and you're also kind and smart. You're also kind of quiet, at least, you are when you don't know people. Nobody knows what's going on in your head, and that intimidates people. There are a ton of guys at school who like you, but they're too scared to talk to you."

Her eyes widened, and her lips parted in surprise as she tried to process Tal's unexpected revelation. "That's crazy."

"You appear pretty perfect. Don't frown. None of this is bad. I just want you to understand. At school, you always seem pretty confident. I didn't realize how many things worry you until we became friends."

"I seem confident?"

"A lot of the time."

"Even in gym?"

He laughed. "Oh no. Never in gym. You're getting better, though."

"I had to ask about gym to see if you are lying to me," she admitted.

"Well, now you know I'm honest."

"That's all really hard to believe."

"Well, it's true. Believe me, I should know. I've been in plenty of conversations with guys talking about how smart you are, and about your hair. I know hair is a shallow thing. Don't hold it against me."

This felt like too much to handle. This was going to mess her concentration up even more than it already was. Were Graham and Tal trying to make her useless?

It felt like Graham had checked a million rooms and he still hadn't found Brake. Why did people have houses this big? Was there a point? It seemed like an awful lot of space to dust, especially when only a few people lived inside.

When he came to the room he'd stayed in his first night in Akkron, the door was open. He peeked inside and saw Brake kneeling on the floor with his head bowed. He was clutching something in his hands.

"Brake?" Graham asked. Brake's head jerked up, and he inspected Graham. Graham felt a pang of concern. Brake was always so composed but now he appeared utterly vulnerable. His warm black eyes were bloodshot and watery with grief. A single tear trickled down Brake's smooth cheek, leaving a glistening trail. Graham hurried to Brake's side, dropping to his knees and placing a hand on his mentor's back in comfort. He could feel the tension in Brake's shoulder. It was a side of Brake he had never witnessed before, and it shook him to the core.

"Brake, what's wrong?" he asked

Brake held up Graham's basketball jersey. "I found this." He wiped his eyes.

"It's my jersey. That's what I was wearing when I came to Akkron."

"Why does it say Dryson?" Brake asked, eyes fixed on Graham.

"That's my last name."

Brake was gazing at him so intently that Graham couldn't help fidgeting. "You said you were raised by your aunt. Did you have an uncle?"

"Yes, but he died when I was little."

"Was his name Lance Baskum?"

Graham's stomach dropped. He had a feeling everything was about to change. Goosebumps rippled over his arms and legs. "Yes," he replied hesitantly, his heart pounding.

"Dryson," Brake whispered, his hands reaching out to touch Graham's face. Tears were overflowing and Graham wasn't sure what to do. "I thought we would never find you again." Brake pulled Graham into a tight embrace.

"Wait, you know who I am? Do you know who my parents are?"

Brake choked back a sob, his grip on Graham tightening so that he could hardly breathe. "I do. You are my son."

Wren opened a portal, and Zera emerged, appearing flawless as usual. Her blue dress brought out her eyes and her long, brown hair glistened in the light.

"Thank you Wren," she smiled. "Brake sent a message saying I needed to get here immediately."

"We haven't seen him yet," Tal informed her. "Graham's upstairs looking for him."

"He told me he was in the room Graham stayed in while he was here."

"Oh, okay. Follow me," Wren said, leading them to the stairs. "Do I need to make portals for anyone else?"

"I do not know," Zera admitted. "He said he discovered a development. I am not sure what it is, though."

"Well, we wanted to talk to you guys anyway, so this is convenient," Tal said, taking the stairs two at a time.

Wren led them down the hallway to the guest room. When they entered, they stopped in their tracks. Brake and Graham were kneeling on the floor, hugging each other and crying. Wren glanced at Tal, but he seemed as bewildered as she was.

Zera approached Brake cautiously, her hand trembling as she reached out and ran her fingers over his hair. He opened his eyes and looked at her. He smiled, the biggest smile Wren had ever seen from him.

"Brake?" Zera asked, her lip quivering. Brake held out one of his hands and took Zera's and nodded, but she pulled away, taking several steps back. Her eyes were wide. "Graham? Right in front of us?" Zera asked, covering her face with her hands.

Brake had one arm around Graham, and he extended his other to Zera. "I suppose we should have asked more questions."

"No, no, Brake," Zera stammered. "I have to know for sure. I cannot do this if it is not true." Tears were running down her face.

Wren had never seen Zera undone like this. It created an anxious knot in her stomach.

"What are we missing?" Tal whispered to Wren. She shrugged, at a loss.

"Graham's uncle was Lance Baskum," Brake said, standing up and holding up the shirt Graham had been wearing the day Wren met him.

Zera softly touched the shirt, and whispered, "Dryson."

Graham stood up slowly and wiped his eyes. His forehead was scrunched, and he was studying Zera like he'd never seen her before. "Does this mean that you're my mom?"

Wren sucked in a breath at the same time Tal muttered, "Whoa."

Zera nodded and held out her arms. Graham wrapped her in his arms and a fresh wave of tears streamed down his face. Zera pressed her face against his shoulder and held on tight. Brake put his arms around both of them.

Wren felt Tal's hand at her elbow and she peered up at him through her tear-filled eyes. He motioned to the door with his head, and they quietly left the room. They walked down the hallway and down the stairs in silence.

"That was crazy," Tal said, running a hand through his wavy brown hair.

"I can't believe Zera's Graham's mom."

"Yeah, this is unexpected."

"Wait ..." Wren said, trying to keep up with the thoughts that were hurling themselves around her brain. "Do you think Brake is his dad?"

"You're only catching that now?" Tal teased.

"Oh wow. This is so big. Brake and Zera? I wonder if my dad knows about all of this."

"I knew Brake and Zera were married."

Wren crossed her arms. "You did? Why didn't you tell us?"

"I assumed it wasn't a great marriage if they didn't live together. I figured it wasn't any of my business. My father mentioned it a few times."

Wren drummed her fingers on her arms. "Do you think this will change things?"

"Probably. We can count on pushing back our trip to Earth."

# CHAPTER 8

"The suspense here is killing me," Ming Li complained from a red cushion in the meetinghouse. "The *one* time I stay behind, something crazy happens and nobody will tell me what."

Graham smiled. It was the third time Ming Li had said the same thing. She could say it a hundred times and Graham would still be smiling. He had parents. Not only did he have parents, they were already people he cared about. He looked at Brake, at his dad, and his heart felt full.

"We'll tell everyone at the same time," Brake said, also for the third time. "That will make it less stressful on everyone."

Graham glanced around the room. Almost everyone was here, sitting in a sloppy circle. Austra, Wren, Ming Li, Sen, Hamble, Brog, and Hedder. Zera was in the kitchen making everyone hot chocolate. The weather had shifted and the meeting place was covered in a fine layer of snow. A small clicking noise sounded, followed by Tal falling into the hideout. He bounced off the large cushion and sat on a chair next to Wren.

"What's that?" she asked, pointing at his hands.

"Oh, uh ... just a little frog."

Wren leaned forward. "I've never seen a blue frog before."

"His name is Saam."

"Hi Saam," Wren said, rubbing her finger over the little frog's back.

Austra shuddered. "At least it's not a rednax this time."

Tal's eyes twinkled. "Nope. Saam's a poisonous frog."

"And you're holding it?" Graham asked, not taking his eyes away from the frog.

"Don't worry. He won't hurt you unless you eat him."

Graham relaxed. "Well, there isn't much chance of that."

"I'm joking. He's not poisonous at all. Li told me you have poisonous blue frogs on Earth. He was cold, so I thought I would warm him up. All the animals are cold. They aren't used to the snow."

Ming Li touched the frog's head. "Saam is a weird name for a frog."

"Would you prefer Sir Croakington the third?"

"That's way worse."

"True. And it's his brother's name, so it would get confusing."

Zera came in with a tray of hot chocolate and placed it on the table in the middle of the room. She handed a cup to Graham and sat beside him, linking her arm with his. It felt so good. Graham wanted to cry. This was what it felt like to have a mom.

The clicking noise came again, and Tram fell into the room. He was a lot slower getting up off the cushion. He was breathing hard from climbing the tree. It wasn't easy to move that much mass around.

"Sorry to be slow," he said, sitting next to Austra. He patted his messy comb over into place. "I didn't see any reason to have a portal when I was already in the area."

"Let's get started," Brake said, with a huge smile. "First off, we want to welcome Sendo to our group." Everyone turned to Sen. He nodded at the curious stares. "For those of you who missed it, Sen made the crystal glow at Meegore." The people who didn't know looked around in surprise. "Sen is from Boztoll."

"This is great news," Hedder said, running his hand over the uneven stubble on his chin. It appeared he was going to try facial hair again. It looked like it was going to be as big of a disaster as last time. "That means things can really start taking off."

"I'm also working on naming this world," Graham said. He couldn't help it. Professor Hedder hated the fact that Graham wanted to name everything.

Hedder rolled his eyes. "I'm sure you are."

Sen seemed confused. "Naming the world? Why? What's wrong with Basura?"

Graham scratched his head. "What's Basura?"

"It's the name of the world."

Everyone stared at Sen.

Hedder narrowed his eyes. "I am a geography teacher and I have never heard that before. I think you're mistaken."

Sen shrugged. "We always called it Basura at my house."

Graham smirked at Hedder. "Then we are calling it Basura. It's about time this place had a name."

Ming Li tilted her head and studied Graham. "Is something wrong with you, Graham?"

"No, everything's perfect."

"I don't take Zera as a touchy person," she said, pointing to their linked arms.

Zera's smile widened, if that were possible. "I cannot hold it in anymore," she said, turning to Austra. "We found out that Graham is Dryson." An audible gasp sounded around the room. Austra's hand flew to her chest, and Hamble's eyes bulged in disbelief. Tram covered his eyes with a trembling hand and locked his eyes on the floor. Only Hedder and Brog seemed unfazed, not making an association with the name.

Ming Li shrugged. "We already knew Graham's last name was Dryson."

"Are you sure?" Austra asked Zera, voice thick with emotion. Zera nodded and Austra's eyes filled with tears.

"What are we missing?" Brog asked from his place on the floor.

"Dryson is my son," Brake said. "Mine and Zera's."

Ming Li sat up taller. "Whaaaaaat?"

"I'm going to tell the story now and you can ask questions later. Agreed?"

Everyone nodded. Graham leaned forward in his chair, ignoring his hot chocolate.

"Most of you know that Zera, Drew, Hamble, Tram, Austra, and a few others used to work for Governor Jorin. Some of you know that Zera and I are married."

"I can't believe you never told me any of this, Uncle Brake," Ming Li scolded.

"There are some things we decided not to remind people of. We were all good at bringing criminals to justice. We tracked some of them to Earth and brought all of them back to Akkron, all except one. Zera and I had enough portal dust to go to Earth and come back two more times. We went to find this last man. His name is Gorbin.

"I took a job at a factory that Gorbin owned in California. He didn't work with the employees and he refused to meet with me. I was working with a man named Lance. We became work friends. Zera and I rented a place, and she spent most of the time taking care of our baby, Dryson. She had been ill for quite a while and was working on regaining her strength.

"We received a message from Magnalee, telling us she needed to meet immediately. She said she had information that was going to turn everything around. In an error of judgment, we left Dryson with my friend, Lance, and his wife. We thought we would be back quickly, but someone stole the last of our portal dust and we couldn't return.

"We searched everywhere for more portal dust, with no luck. I tried bribing every goblin I could find, but there wasn't any left. When our group disbanded, Zera and I decided we also needed to go our separate ways, so we could work without suspicion. We kept fighting for good, even though our hearts were broken."

Graham swallowed a lump in his throat. His eyes were burning with unshed tears.

"Years later, I was finally able to come up with some portal dust. There was only enough to go there and come back. I went, and I searched for Dryson, and I looked for Gorbin. Dryson and Lance were nowhere to be found. I asked everyone I could find that worked for Gorbin, but no one had ever heard of Lance. It had been too long, and he wasn't in the house we had left him at.

"I chased after Gorbin. I admit, I should have waited. Going in angry and tired didn't help me. I found him and demanded to know what happened to Lance. He didn't know who I was, and he didn't know Lance. I tried to throw a fireball at him, but he blocked me.

"When he realized I had magic, we fought. I don't know how long it lasted. I must have looked pretty bad, because he left me for dead. I thought I was done for until Ming Li and Mali found me. My head was foggy, and I used the last of my dust to take us back to Akkron. I've never forgiven myself."

"I never blamed you," Zera said, taking Brake's hand. "You did the best you could."

Graham glanced around. Austra, Hamble, and Wren were drying their eyes and Brog, Hedder, and Sen appeared serious. Ming Li's eyes were bouncing around like she was thinking, and Tram was still staring at the floor.

"Wait," Graham said, jiggling his leg. "My aunt told me that my parents were on a cruise."

"We told them that was what we were doing," Zera explained.

"But the boat capsized and there were witnesses who said they saw my parents on it."

"I don't understand that," Brake said, rubbing his chin.

"Neither do I," Zera said, scrunching up her forehead.

Tram shifted nervously in his chair and raised his hand. "I was the cleanup guy."

"What does that mean?" Tal asked, stroking Saam's head.

"It was always Tram's job to make sure there weren't any traces of us being anywhere," Hamble told them.

"They left me behind when Zera and Brake went to talk to Magnalee. When they didn't come back, I realized something was wrong, so a decision had to be made. I knew you weren't coming back, so when I heard a boat had capsized, I hurried and paid off a few witnesses to say that you were on the boat. I figured that way Dryson could stay with those people and he would be taken care of."

"Why didn't you bring him back with you?" Zera asked, narrowing her eyes.

Tram wiped sweat from his round face. "I couldn't. You know I've never been able to make a good portal, even with the dust. I figured it was too dangerous to go through carrying a baby."

"That is pretty pathetic," Sen said, shaking his head. "If you could go through, it shouldn't have been hard to go through carrying something."

He took out a handkerchief and wiped his brow. "Well, I had to carry all the gold we left there. If I didn't bring it back, it would have been a significant loss to our mission."

Zera stood up and walked over to Tram and smacked him across the face. "You chose money over my baby?"

Tram rubbed his face and gazed at the floor. "We needed that gold."

"Why? We all have tons of money!" Zera was breathing hard and Graham was afraid she was going to do something she might regret.

"*You* all had money," Tram said, crossing his arms. "I never did."

"We would have rewarded you with anything if you had brought Dryson back."

"You say that now," Tram said, puffing out his chest. "But you would have just taken the baby and said thank you."

"You are almost as bad as Fria."

Graham stood up and hurried over to his mom. "It's okay, mom. We're together now. Let's be happy and forget about what we missed. Please?"

Zera studied him and nodded.

"The gold from that mission never actually came back," Hamble said. "Brog? Would you mind escorting Tram to the prison?"

"Happy to," the giant said, standing. He pulled Tram to his feet and led him out the door. He didn't put up a fight.

"I'll go with him," Hedder said, standing. "Just to make sure he doesn't try anything."

"No more secrets," Brake said. "And when we talk about the things we are doing, we give all the details. I can't help feeling like I should have questioned Graham more in the beginning. If I had, I might have found out he was my son sooner."

"I'm kind of glad we didn't find out right away," Graham admitted. "You were both kind to me before you knew I was your son. I already feel a connection to you. It would have been more awkward to find out when I didn't know you." He peeked at Zera and her eyes gleamed.

Brake nodded. "Be that as it may, we don't want to make any mistakes because we haven't shared information. Raise your hand if you agree." Everyone's hands raised. "Great. Does anyone have anything they need to tell us? It could be big or small."

Hamble rubbed his bald head. "I wouldn't normally think you meant relationship information, but since that brought this whole conversation up, do we want to talk about those some more?"

"I'm not sure I understand what you mean."

"It's probably not important, but if we are going to have the past involved in the future, do we need everyone to understand all of our past relationships? Like Graham might have guessed you were his parents if he knew you were married and that you had a son that was lost on Earth."

"We do not have any other relationships like that."

"Well, what about things like arranged marriages–"

"I don't see how that would matter right now," Zera cut in.

Zera answered too fast. Graham wasn't sure if he wanted to know if he had an arranged marriage. Or maybe Hamble meant Brake and Zera.

"Did you have an arranged marriage?" he asked his mom.

She smiled, and her eyes twinkled. "We did."

"You did?"

Brake chuckled. "Arranged by Zera."

"Brake and I met in school. We were assigned to be partners on a project. He was so smart and handsome. I begged my father to talk to his parents and arrange something."

Brake winked at Zera. "To say Zera was spoiled is putting it mildly. Her father had it all arranged by the end of the day."

"Were you okay with it?" Wren asked.

"Of course," he winked again. "Have you ever seen Zera in action? She can do anything she sets her mind to. I knew I was the luckiest man alive."

"It all makes sense," Ming Li said. "From what I hear, Graham's medicine tastes as bad as Brake's cooking, and he would have to have good looking parents because check him out." Graham groaned and everyone else laughed.

"They were lucky to have what they did," Hamble said. "Everyone isn't as lucky. I always felt bad for Drew and Magnalee. They would have been better off as friends."

Graham glanced at Wren and watched as her eyes widened with understanding. Austra poked Hamble and everything was quiet for a moment.

"My parents had an arranged marriage?" she asked. "I should have guessed. I wonder why my dad didn't tell me."

"Everyone has to make the decisions they think are best," Zera said. "I believe it would have all worked out in the end, if Magnalee had not died."

Wren nodded. Graham couldn't tell what she was thinking.

"I don't understand why the governor chose you guys to fight criminals," Graham said. "If I was a baby, you all would have been really young when it started."

"You don't believe young people can be chosen to do big things?" Brake asked with a small smile.

"It was because we defeated a sorcerer," Hamble piped in.

"A sorcerer?" Graham felt confused. "Isn't a sorcerer someone who does magic? Doesn't that make everyone a sorcerer?"

"We only call someone a sorcerer if they study magic and can do more than the normal person," Hamble explained. "They are also almost always evil. People who spend all of their time trying to have more power are rarely up to any good."

"Zera, Fria, and I were in a magical equality class that Hamble taught," Brake explained.

Ming Li snorted. "Hamble was a teacher?"

"I was an outstanding teacher," he said, side-eyeing Ming Li. "There was a sorcerer that was causing a lot of trouble for the trolls. It came to my attention, and I wondered if I could help. I asked Tram to help. He was a history teacher. We made a plan and some of my snoopy students found out I was up to something and insisted on joining."

"I wouldn't call us snoopy," Brake protested.

Zera smirked. "We were snoopy."

"Drew had graduated the year before, and Brake convinced him to join," Hamble said. "Austra was new to town and had earned the reputation of being a dangerous beauty."

"My goodness," Austra muttered, as she rolled her eyes.

"We heard a rumor Austra was planning on chasing after the sorcerer, so we asked her to join us," Hamble explained. "I'm not going to tell the entire story, but we defeated the sorcerer, and that won us a lot of respect. That was why the governor chose us."

Tal leaned forward. "So back to arranged marriages. Does Graham have one?"

Zera shook her head. "No."

"Darn."

"What?" Graham said, grinning mischievously at Tal, "Misery loves company?"

"Wait what?" Ming Li piped in. "Tal has an arranged marriage?"

Tal glared at Graham. "I'm working on getting out of it."

Ming Li giggled and turned to Wren. "Well, that simplifies some things. Now you can choose Graham."

Graham coughed and watched Wren's face go from pale to red in record time. Her eyes narrowed, and she gritted her teeth. She was angry. Really angry. Graham wouldn't have been surprised if she started breathing fire. Ming Li must have noticed because she shot up and said something about looking for something in the kitchen and quickly disappeared. Wren jumped up and followed her out the door.

Tal scanned the room. "Ummm ... do you think we should follow? Just to make sure Wren doesn't kill Li?"

Hamble shook his head. "No siree. Never follow a person with that expression on their face. I haven't seen a look like that since Drew told Austra he wasn't going to mmrph–"

Graham's eyes widened as he watched Austra throw Hamble to the floor and put him in a headlock, covering his mouth with her hand.

She growled. "Not. Another. Word."

Brake smacked his hand on his leg and threw his head back and laughed. "This is just like old times."

"Except you just had one of your friends sent to prison," Sen stated. "No one seems concerned about it either."

The room was silent for a moment, and then Hamble shrugged. "Well, we've sent him to prison a time or two before. This time it will be a little more permanent, is all."

Austra sniffed. "He always was rather annoying and self-serving."

"I have a question," Tal said. "How did Hedder get involved in all of this? He's too young to be part of your group from when you worked with the other governor."

"Hedder had it rough growing up," Brake explained. "He never fit in at school and his parents were strict. After he graduated, someone from The Dark Cloud tried to recruit him. They bothered him for months and didn't want to give up on him. He remembered the stories about us and he came to me for help. Zera was able to get him a job at the school and made it known she was supporting him. The Dark Cloud must have realized he wasn't going to follow them, and they left him alone."

"Speaking of Professor Hedder," Zera said, "He has agreed to tutor you all in history and geography, as well as math."

"Aw man," Tal moaned, "What did we do to deserve that?"

Zera tilted her head as she studied him. "It is not good that we have allowed your education to slip. I will teach you the subjects Professor Hedder cannot. We need to make sure you do not fall behind."

Graham sighed. Nothing sounded worse than spending that much time with Hedder, especially now that he had parents.

"Any more secrets?" Hamble asked.

Zera cleared her throat. "We better dismiss this meeting. I do not think we can handle any more secrets coming to light today."

"Why are you always embarrassing me?" Wren asked as she watched Ming Li chug a glass of water.

Ming Li put her glass down and turned to face her. She fiddled with her fingers and then stared at the ceiling. "I don't mean to embarrass you. I don't really think before I talk, and I don't think about how it might make other people feel. I know it's a personality flaw, but I don't know if I can stop."

"You can stop. Just pause and think before you say something," Wren said, plopping onto a kitchen chair.

"I would have to think to pause, and I'm not sure I can make that into an automatic thing."

"If it involves me and boys, don't say it."

Ming Li sat beside her. "If I leave your love life to you, nothing will ever happen. Do you really want to lose Graham because you never did anything?"

"Lose Graham? I don't have Graham." Sometimes Wren wondered if it was even worth trying to make Ming Li understand the way she felt. "I also don't *need* a love life right now."

"You won't ever get him if you don't try. I thought I was easing your mind by telling you that you don't need to think about ever liking Tal. If he has an arranged marriage, that makes some things a lot easier."

"But did you really have to say it in front of everyone?"

"It wasn't everyone."

"Everyone I wouldn't want you to say it in front of, especially Graham."

Ming Li didn't make eye contact with her, "I already told you he likes you, so why should it matter?"

"You told me he likes me," Wren said, leaning her head against the chair and crossing her arms. "That doesn't mean he does. And you said it in front of the adults."

"They're all used to me," Ming Li shrugged. "Nobody takes anything I say seriously. Wait. Did you decide you actually like Tal? Is that why you're upset? There isn't much chance of getting out of an arranged marriage."

Wren shook her head. "No, I don't like Tal. Not like that." This was a waste of time. "Just forget it, okay? We have a lot to plan without worrying about boys. Just try not to embarrass me. Please?"

"I'll try," Ming Li said, tilting her head and shrugging. "I can't promise anything, though. You know I do it to everyone."

"That doesn't make it alright. It's really rude."

"You can punch me every time I do it," Ming Li said. "That might help me remember for the next time."

Wren shook her head. "Just watch what you say. Can we change the subject?"

"I still can't believe Brake and Zera are Graham's parents. Should we start calling him Dryson?"

"I don't know," Wren said, glad to be on a different subject. "It depends on what Graham and his parents want. It would be weird to change your name at this age. It might be hard to get used to."

"Yeah, but he was already using Dryson as his last name, so it probably wouldn't be that weird."

"It's still weird to think about having two names."

"It's actually really convenient," Ming Li said. "And you know, someday Graham is going to run for governor or something and he's going to name everything. I bet he even makes last names a thing."

# CHAPTER 9

Wren watched Sen say goodbye to his family. She held her cloak tight. The jungle was still colder than normal. Adding Sen to their group had been interesting. It had only been a few weeks since he joined them and he still didn't talk much, so it was easy to forget he was there. Despite this, Ming Li tried to engage him every chance she could, but he seemed content to disappear into the jungle for long periods of time. The only time he showed any genuine enthusiasm was during competitions, which, he told them, were common occurrences in his family given that he had five brothers and they spent most of their time trying to one up each other.

A few days ago Sen had asked Wren to send his family to Earth. It'd surprised everyone at first because he could do the cool teleporting thing, but it turned out he could only do it on Basura. He couldn't go to different worlds. Sen's father had wanted to return home for years, but had never found a way back. She wondered if his brothers would like it. They had never been to Earth, and without magic, they wouldn't be able to speak the language.

Sen hugged each of his five brothers, who ranged in age from about six to seventeen. He turned to his father, Rosendo, both looking uncertain. Sen's father was a large man with a black mustache and an enormous hat. It was the biggest hat Wren had ever seen. She wondered if he made it himself. It appeared to be made of straw and stuck out at least a foot all the way around.

"Are you sure you want to leave?" Sen asked, fiddling with the edge of his worn black cape.

"Positive mi hijo," Rosendo said, patting Sen's shoulder. "I know you can't come with us, but perhaps the girl can send you to us when your destiny is complete."

"What if you don't like it as much as you think you remember?" Sen's little brother piped in.

"That's my fear," Sen's mother, Kraya, said, biting her lower lip. She tossed her neat blond braid over her shoulder. "You said I'm going to stand out like a pink chicken in Mexico, and I hate attention."

Rosendo drew his wife in for a hug. "I promise I will bring you back if you hate it."

Wren's eyebrows came together, and she tilted her head. "How would you bring her back?"

"Yes, how?" Sen's oldest brother asked, folding his arms and frowning. "You never told us how you came to Basura."

Rosendo sighed. "Is that really important, Miguel?" All six of his sons nodded. He took off his hat and tossed it to the side, sitting down on the dirt. His sons all sat in a circle around him and stared up expectantly. Wren and Kraya both sat on rocks.

"I was a curious teenager," Rosendo said, resting his elbows on his lap. "I had two friends that I got into trouble with regularly. A rumor was going around about a witch named Slosha that lived outside our village. People said that she could fly and cast spells. They claimed she lived far from people so that her secrets wouldn't be discovered. My friends and I were intrigued and wanted to see if it could be true. We decided to go and spy on her.

"We couldn't ask where the witch lived because we didn't want anyone to suspect that we might go searching for her. It took us over a month before we found her small home. It became an adventure for us. On the weekend, we would spy on her from the trees. Slosha didn't look like any witch we had ever envisioned. She had long, silky, black hair, and she wasn't that much older than we were. Twenty-five at the most.

"It didn't take long before we concluded she wasn't a witch. We watched her come and go and do chores around her small farm. Nothing exciting ever happened, but we kept coming anyway."

"She must not have been an ugly witch then," Kraya said, with a twinkle in her eye.

Rosendo winked at his wife. "You already know this story, mi amor. No, the witch wasn't ugly, but she was not as beautiful as you."

"Gross, Papá!" Sen's chubby little brother said, covering his face with his hands. Wren smiled inwardly at the boy's embarrassment. "Just tell the story of how you got here."

"One day, we decided it was time to move on to another adventure. We would spy on Slosha one last time. We positioned ourselves behind our favorite trees and half heartedly watched. Slosha was outside working in her garden, pulling weeds. Thunder sounded in the distance and she stood up and peered up into the sky. She snapped her fingers and was suddenly holding a shovel."

"So she was a witch!" the youngest brother exclaimed, bouncing up and down. "Did she send you here? What happened?"

Rosendo shook his head. "We couldn't believe our eyes. We were too scared to do anything that day, so we returned home and made a plan. The next week, we went to her door like a bunch of fools. When she answered, we told her we knew she was a witch. Her eyes were wide with fear and she let us in. We told her we would keep her secret if she did magic for us whenever we needed it."

"You threatened her?" Sen asked, his mouth turning down.

"I'm not proud of our actions," Rosendo said, peering at the dirt. "She told us about a waterfall that had magical powers. She shared the location with us, and it was a place we knew well. It was a popular place to swim in the summer. She said if we jumped off the small falls and yelled a certain word as we descended, we would have our own magic.

"We felt that having our own magic would be better than using Slosha's, and so we proceeded to the waterfall immediately. We climbed to the top, and all jumped from the falls, yelling the word. The next thing we knew, we were smashing into the dry ground. When we got up, there was no water in sight. We were in a forest surrounded by trees. We had been transported to a place just outside Akkron."

Wren leaned forward. "You must have gone through a permanent portal. The word she had you say must have been the trigger word to open it."

"Perhaps," Rosendo said, with a far off look in his eye. "We tried to get home for ages. Trying to ask for help was pointless and everyone treated us like we were loco. We couldn't speak the language, so we had to draw pictures. We eventually gave up and went searching for work. I found a job on a farm. My friends didn't want to farm, so they traveled elsewhere. I never saw them again."

"What next? What next?" the youngest boy said with a grin.

Kraya's eyes sparkled. "The farmer had a fantastic daughter that fell in love with your father and taught him the language. He taught her his language, and they eventually married and moved to Boztoll and had six handsome sons."

"Six?" Miguel, the oldest, sniffed. "I would say three, possibly four."

"And you are not one of those three," Sen said, pushing his shoulder into his brother.

"Please," Miguel grinned, "I am the most handsome of the six of us."

"You are all handsome, and I think it's time to go," Kraya said, standing up and brushing the dirt off her blue dress. She pulled Sen to his feet and wrapped him in a tight hug. She whispered something Wren couldn't hear into his ear, and he nodded.

Rosendo rose slowly to his feet, and the others followed. "If you hate it, we can jump back through the portal at the waterfall. Deal?" he held out his hand and Kraya took it, nodding. "Leave your capes here, boys," he said. "People will think we are really weird if we wear them on Earth." They all dropped their capes on the ground and looked at Wren expectantly.

"Come when you can," Kraya said, cupping Sen's face in her hands. He nodded.

Wren opened a portal and hoped the map Rosendo made was accurate enough to get them to the right place. "I wonder if I should go through with you and make sure it's the right place."

"Thank you," Kraya said. "I was nervous."

"Watch your step as you go through. It's a bit of a drop." Wren watched as Sen's brothers leaped through the portal and fell on the other side. She offered Rosendo and Kraya her hands, and they stepped through without a problem. It was humid. They were standing on a dirt road surrounded by large green shrubbery. Up ahead she could see a small village. "Does it seem familiar?"

"Yes," Rosendo beamed, hugging Kraya with one arm. "We are close to my village. We can be there in ten minutes. Thank you."

"Can you remember this place?" Kraya asked her. "Will you be able to send Sen one day?"

"Yes," Wren said, "But will he be able to find you in the village?"

"I didn't think of that," Rosendo muttered. "Tell Sen I will come back in one year to see if he is ready."

"I will." Wren said, giving them all a small wave. She jumped back through the portal. Sen was staring blankly at the shimmering portal. It was hard to gauge how he felt about his family going to live in a different world. He'd told them the night before that it would be a good change for them. Now they wouldn't have to stay in Boztoll with The Dark Cloud hovering over them.

"Are you going to be alright?" Wren asked him as the portal closed.

"Yep," he said, turning and walking towards the meetinghouse tree.

Wren hoped he was telling the truth. She sat back on the rock and yawned. Tal's frog Saam hopped up to her foot, and she scooped him up. The little frog had become needy since Tal took him in, insisting on sitting on someone's lap at all times while getting his head stroked. Despite it being three weeks since the last snow, it was still a little too cold for all the animals. Sen had been busy making warm spots for them to gather.

She couldn't believe it had been three weeks since they found out Brake and Zera were Graham's parents and they had gotten little done since. They told Zera and Brake about their plan to go to Earth and capture Gorbin, and that had received a ton of backlash. All the adults felt it would be a waste of time and that the focus right now needed to be on figuring out the silver eclipse. After a lot of debate, they had agreed; Gorbin was only one person, and he didn't seem to be a threat at the moment.

Zera and Brake had been spending as much time as they could with Graham at the meetinghouse. It was slowing their plans, but Wren could understand why they needed to get to know each other as a family. She would be thrilled if she had the chance to be with her mom.

"Ahem," a small voice cleared its voice to the side of her. Wren jumped up, nearly dropping Saam, and looked down to see a small ... creature. She wasn't sure what it was. It was about a foot tall. Its dark blue body was complemented by a light orange back and unusual swirls in a lighter blue shade. It had a large mallet shaped nose that hung over its frown. Its body was round, with stubby wings and arms that appeared too small to be useful. The creature's head grew exactly one thick curled hair. It stood upright on webbed feet.

"Can I help you?" she asked, staring down at it.

"I am here to help you," he said, in a high, almost shrill voice. "And I could have helped you sooner if you were easier to find." He placed his hands on what was probably his hips and glared up at her. "I've been searching high and low for you these past three weeks. You and your friends were in and out of the cave like your britches were on fire. Life is better appreciated if you take time to enjoy your surroundings."

"Who are you?"

"Who am I? Nobody sees you in a few hundred years and they forget you, I suppose. I am Padmire. I'm a bungle. The only bungle left, in fact."

Wren bit her lip. "What's a bungle?"

"I am!" he snorted. He held up his arms and turned in a slow circle, as if that would help her know what he was. "You've never heard of me?"

Wren shifted her weight from side to side. "Sorry."

"Kids," he snorted again. "They never study their history. I thought you were the smart one."

Wren peered around the forest, wondering if this was some sort of trap, but it all seemed peaceful.

"I was created to care for the magic at Meegore. My creators were not as talented as they thought they were and bungled us up pretty good. I was the only one who survived for more than a week. They wanted me to fly, but they made my wings too small and my body too heavy. They gave me webbed feet to swim, but silly little arms. I could have contacted you the first time you entered the cave, but when the brown-haired boy kicked me in the water, I went whirling away and couldn't catch up because of these ridiculous arms."

Wren's eyes widened in realization. "So, you were the creature that touched Tal's leg?"

"Yes, and instead of waiting to see if I was alright, you went tromping off."

"We thought you were a fish or a snake or something."

"Typical."

"I'm sorry. We would have stopped if we had known."

"Well, now I've been trudging all over creation, searching for you. Your scents are easy enough to follow, but these silly feet make it difficult." He held up a webbed foot and wiggled it. "You are not the easiest people to help. I left *The Silver Eclipse* book right out in the open and you didn't even take it! I couldn't bring it with me because it was too heavy and I only have magic inside the cave."

"Wait, are you the one who made me get stuck?"

"Of course I was. I can make the cave do anything I want."

Wren's eyes narrowed. "Why did you do that? I was terrified!"

"I thought if I could hold you in one place, your friends would all come and I could talk to you all at once. Unfortunately, only one friend slid down the slide and the others were all being useless up above. I released you and climbed along the ceiling, waiting for an opportune moment to make my presence known."

"So why didn't you?"

"The black-haired girl and the new boy were going too fast, and the conversation I had to endure from the other two boys was painful, to say the least. Coming in at that moment would have been much too awkward, and I do not do awkward. I gave you all a minute in the final cavern, and before I knew what happened, you had already left."

Wren cocked her head. "What help do you have for us?" She still wasn't sure if she trusted him.

"Bring your friends out here and I will tell you. I refuse to go into a place that is there, but doesn't appear to be there."

Wren shrugged. That sounded like the cave to her. "Wait here and I'll be right back." She climbed the tree as fast as she could and fell backwards over the secret branch and into the meetinghouse. Her heart was pumping fast. She hoped whatever Padmire had to tell them would be something they had been waiting for.

Graham studied the strange blue bungle and decided not to get too close. It seemed angry, and it had sharp teeth. It had surprised them when Wren came bursting into the meetinghouse and told them about the creature. Now the five of them were all outside, waiting to see what it had to say.

"Can you all come to my level please?" the bungle squeaked. "The view I have up your noses is distracting."

Tal snorted, and they all sat facing Padmire.

"What I have to say is secret," he said, glaring at each one of them. "I will only tell you if you promise to keep it that way. Do you promise?"

"I promise," Ming Li, Wren, Graham, and Tal all said in unison. Sen nodded.

"Say it, new boy."

Sen's mouth turned down. "I promise. And my name is Sen."

"I have been waiting for you inside the cave for so many years. Before the cave was closed, many people would come to receive more magic or to get magic. My job filled me with happiness. People came in and were so excited. I was part of something great. It is my privilege to choose what magic people receive.

"I can sense what people need and I used that sense to help people. I guided them to the magic that would best suit them. After some time, I saw people change. People that I respected were becoming overconfident and unfeeling. All they cared about was more power.

"One group used their magic to try to take over the continent. This couldn't happen. I had the crystal taken from inside the final cavern and placed outside the cave. I sealed the cave until the five would come, when the world needed them. No more magic for those seeking for power."

Tal leaned forward. "So you've been alone inside the cave until now?"

"Yes."

"That must have been lonely," Wren said.

"I am good at amusing myself. I was happy when the cave opened again. It's good to be useful."

"So you chose what magic we all got?" Graham asked, thinking about the floating bubbles.

Padmire puffed out his chest. "Of course I did."

Wren tilted her head and seemed to be trying to read the bungle. "So the color of the bubbles didn't matter?"

He grinned from ear to ear, showing his uneven, jagged teeth. "Not at all. In fact, the bubbles don't even matter."

"Then why have them?"

"Because people need something to hold. You are an untrusting species and you need proof of anything. The bubble helps people believe they are getting something."

Ming Li fiddled with her braid. "So if you give people the magic, couldn't you have stopped and not helped the greedy people?"

"Unfortunately not. If someone gets through the cave, they will get magic. I can only guide it. If I do nothing, it will be random."

Tal shook his head. "We didn't get any new magic when we went through the last time."

Padmire rolled his beady black eyes. "I always wonder why children are chosen. You are so untrained. The first time you passed through was a shock to your system. It was a feeling you didn't know, so it came out strongly. The more you enter the cave, the harder it is to figure out what you have been given."

"It was Sen's first time through, and he doesn't know what his is," Ming Li argued.

"I actually do," Sen said. "I can throw fire."

"WHAT!" Ming Li yelled, "And you didn't tell us?"

"I didn't know it was a tell all."

"When did you find out?"

"The night we came here. I haven't practiced because, you know, fire. I don't think it would be practical to use in the jungle."

Padmire nodded and turned to Wren. "Girl with long, fiery locks. Do you know what you have?"

Wren pushed a strand of hair behind her ear. "No, and I'm Wren."

"I prefer to call people by their hair color."

Ming Li shook her head. "That's stupid. Tons of people have the same color hair. Graham and Sen, for instance."

"That is why I called one the new boy. As he isn't new anymore, I can call him the boy with black, straight hair and the other one can be the boy with black, curly hair."

Graham laughed. "Did you create this world? It is so much easier to call people by their names. That's something I want to change around here. Everything should have a name, not just a description."

"I'll think about it," Padmire huffed. "Now red hair. Wren. Are you telling me you have no ideas as to what your new magic is?"

"None."

"Open up the sky," he said, nodding at Tal.

Tal's eyebrows bunched together. "Huh?"

"You are the plant boy. Part the trees."

"Oh ... okay," Tal said, standing. He looked up and spread his arms out wide. The branches of the trees moved out of the way, creating a circle of light.

"Good, good," Padmire squeaked. "Now red ... Wren. You see the cloud above us?"

"Yes."

"Stand up."

Wren stood and crossed her arms, fiddling with her blue sleeve. "What am I supposed to be doing?"

"Stare up at the cloud and raise one arm in the air. Point at it."

Wren stretched her arm out and pointed. Graham could tell she felt ridiculous.

"Now, connect to the cloud. Put all of your focus on the cloud."

Wren narrowed her eyes. "I feel something."

"It is only you and the cloud. Focus."

"Is her hair supposed to do that?" Ming Li asked, pointing at Wren. Her red hair was standing on end, making her look like a red porcupine.

Tal raised his brow. "That's some serious static."

"Now bring your arm swiftly towards that tree," Padmire said, pointing.

Wren brought her arm down and lightning brighter than anything Graham had ever seen struck the tree, followed by a boom of thunder. They all fell to the ground, and the tree caught fire, rapidly burning.

"Not the tree!" Tal yelled, as Graham jumped up and shot ice up into its branches.

"If that spreads, the whole jungle will be gone!" Ming Li called out.

"Graham, warm up your ice," Padmire said calmly.

"What?" Graham asked, searching around for anything helpful.

"Don't shoot ice, shoot water."

Graham thought warm thoughts as he flung more ice. After what felt like an eternity, water streamed from his hands and into the tree. The fire simmered and went out. Graham put his hands to his legs and breathed deeply. He wasn't sure if it was water or sweat rolling down his face.

Tal sped to the tree and put his hands on its trunk. "Why did you have her do that?"

"It will grow back stronger."

"That doesn't help right now," Tal said, putting his hands into the air and twisting them around each other. The dead parts of the tree fell off and new green growth appeared. "There. Don't do that again," he scolded, glaring at Wren.

"Sorry," she said, throwing her arms into the air. "I didn't know that would happen!"

"So, what can I do?" Ming Li asked.

"You can levitate. That is useful to anyone, and most people can already do it to some degree."

"And Tal?"

"He can summon objects ... even if he doesn't know exactly where they are. Of course, they have to exist, and you can't summon anything out of places that are protected."

Tal's eyes lit up. "I've never been able to summon," he said, blowing into his palm. A cookie appeared in his hand and he stuffed it into his mouth.

"I've tried to give you gifts that will help you in your quest to produce the silver eclipse. It would have been helpful if you had all brought the book with you."

"What does it say?" Graham asked.

"Do I look like I can read?" Padmire sniffed. "I am called a bungle for a reason. Everything about me is a little messy."

"So you can't help us with the silver eclipse?"

"I helped you. I gave you the magic you need, *and* I presented you with the book. It's not my fault you didn't take it."

"Should we go back and get it?"

"No, you have already wasted enough time. It's time for you to figure it all out."

Graham looked around at his friends and nodded. It was time.

"Do we want to know?" Zera asked, as they all entered the kitchen. Wren's hair was still full of static and she couldn't make it come back down. Graham was dripping wet, and Tal had charred tree pieces on his shoulder.

"We're stepping things up with the silver eclipse," Ming Li said. "We figure we should eat first."

Zera said nothing as she handed Graham a dish towel and put a plate of sandwiches on the table. Wren took one and collapsed into a chair. She had hoped she'd get some sort of magic that would help in a fight, and now she had. Lightning. She figured this was one of those 'be careful what you wish for' moments. Sure, lightning could be useful in a fight, but it seemed pretty violent and permanent. She didn't want to kill anyone, so what good was that type of magic?

Wren studied Sendo. She wondered if he felt the same way. His stoic demeanor and straight face gave nothing away. Throwing fire was almost as bad as hitting someone with lightning. Why hadn't she gotten something like Ming Li? She could stop people with the wind and it didn't hurt anyone. Graham's ice might hurt, but it wasn't like this.

"What is wrong Wren?" Zera asked, sitting next to her.

She swallowed hard. "I can call down lightning and I started a fire."

"In the jungle?" Zera asked with wide eyes.

"Yes, but Graham put it out."

"That must have been scary."

Wren nodded. "I don't want that kind of magic."

"Why?" Tal asked. "I think it's awesome."

"It's not my goal to kill people, even The Dark Cloud."

"You don't have to kill them," Tal said, biting his sandwich.

"Yeah, you can hit near them," Graham said. "You didn't hit any of us and we all fell over."

"That's true," Wren said, relief flooding her worried mind. Knocking over her enemies could be helpful.

"A warning would be nice though," Ming Li added. "That was super loud and bright."

"So how do we do better with the silver eclipse?" Sen asked. "We've been trying to figure it out for weeks."

"It's possible we're looking at too much," Ming Li said. "When I read it, it feels so overwhelming I can't even concentrate. We should take one step at a time and figure it out and ignore everything else. My mom used to tell me to only do what you can. When my room was a mess and I felt like I couldn't clean it, she would tell me not to clean the whole thing. Even picking up one thing gets you closer to your goal. If you pick up one thing every time you go in, it eventually gets done."

"That makes sense," Tal said, tapping the table with his fingers.

"Have you ever cleaned your own room?" Wren asked with a crooked smile.

"I'm sure I have," Tal said, grinning. "Even servants get the day off."

"What was the first step?" Graham asked, trying to remember.

"A hair from a traitor who has a hateful heart," Sen said, finishing his sandwich.

"I feel like this one should be easy," Graham said. "There are probably lots of traitors."

"But it can't be someone like Jaaz," Wren said, remembering the boy who had tried to give them over to The Dark Cloud. "Even though he betrayed us, I don't believe he has a hateful heart."

"That's true."

"I am not sure if we were right to leave the entire silver eclipse quest up to you," Zera said, tapping her chin. "Perhaps we would have insights you do not."

"Do you know someone who fits this description?" Wren asked.

Zera nodded. "The first person who comes to my mind is Fria."

"The one who killed my mom?" Wren asked.

"Yes," Zera said, looking at Graham. "It is difficult to send you all out after someone like her."

"Didn't Austra say you had all been searching for her for like fourteen or fifteen years or something?" Graham said, shifting in his seat. "If you can't find her and you worked with her, what are the chances of us finding her?"

Zera looked at her hands and sighed. "I might know where she is."

"Then why haven't you told anyone?" Wren asked, frowning. "She killed my mom."

"Yes, I know," Zera said, locking eyes with Wren. "Magnalee was my best friend and I will always miss her, but vengeance is an ugly thing. It seems likely Fria would be on Earth with Gorbin. I do not know for sure, and for a long time we could not go to Earth, so dwelling on it was a waste of time."

Wren glanced at her friends. "So, it looks like we are going to Earth after all."

# CHAPTER 10

Three days later, Wren was standing in a place that Graham called a hotel, peering out the window at an enormous city.

"There're so many people," Tal said, watching the people scurrying around below. "It's no wonder Gorbin came to Earth. There are way more people to boss around. It's good my father doesn't know."

"You all know Basura means garbage in Spanish, right?" Ming Li asked. "It took me a bit to remember, and now I think it's horrifying we're calling our world garbage."

They all stared at Sen, who shrugged. "I told you I don't know Spanish. My dad always called it Basura, so we did. He hated it there, so I guess that's why."

"Well, it's too late to change it now," Tal said. "We even have the adults saying it."

"The air is gross here," Sen said, turning away from the window. "I could taste it when we were out there."

"I could never live here," Wren admitted. "Those vehicle things are way too scary, and they're everywhere! That thing we took here was driven by a madman!"

Tal grinned from ear to ear. "I loved that. That taxi thing was great. I hope we can do it again."

Wren shuddered. She was sure the person driving the taxi had less driving experience than Graham. He had been all over the place. Graham claimed people had to pass a test to drive, but Wren wanted to see that guy's proof. It had taken five minutes after they got out for her heart to beat at a normal pace.

"So, how do I look?" Graham asked, coming out of the bathroom. He was wearing a weird black jacket that he called a suit, on top of a white shirt with lots of buttons. He had a green thing around his neck that he assured them was normal when you wanted to appear professional.

"You look ... really weird," Tal said. Wren laughed and Ming Li shook her head.

"Don't listen to Tal," Ming Li said. "You look hot."

"I agree with Tal," Wren said, smiling, "But you don't look as funny as he does."

"Why are you guys always hating on my Earth style?" Tal asked, holding out his arms and turning in a circle so they could examine his outfit.

"Why can't you ever get normal pants?" Graham asked. "I mean pajama bottoms with pigs on them is better than your flamingo swimming suit, but only a little."

"Pigs are awesome and so are pajamas. I bought like ten of them. They are sooo comfortable. I want to wear them forever. I bought some with an animal that has a long neck.

There is no way that is a real animal. I don't think it would be able to walk."

"It's called a giraffe," Ming Li said. "They are real."

"I need to see one."

"We could go to the zoo."

"That's probably not a good idea," Graham told them. "You should probably stay here until I get back."

"That's so boring," Tal argued. "You said this thing you're going to lasts three days. You really expect us to stay in here that long?"

"Ming Li can show you the T.V. I'm sure you'll be entertained for hours."

Wren was happy to stay here. This world seemed a little overwhelming to her. She still couldn't believe how lucky they had been to come right when Gorbin Corporation was holding a conference for potential employees. Graham was able to get one of the last spots. The only jobs he could actually apply for were cleaning positions and the cafeteria, but it wasn't like he was going to work there, anyway.

"Just out of curiosity," Tal said, "If you happen to see Fria, how are you going to get a hair? It's kind of weird to walk up to someone and yank a piece out."

"I've been around Wren and Ming Li enough to hope that if I can get behind her, she'll have some stray hairs on her back or something."

"Wait ... what are you saying?" Wren gave Graham her best glare.

Tal chuckled. "He's saying you two shed like crazy. I had to unclog the sink in the meetinghouse and it was full of long red and black hair. There was probably enough to make a toupee."

"Do you think Fria will even be going by her real name?" Ming Li asked.

Graham shrugged. "I hope so. Gorbin's using his real name, so there's probably a good chance."

"You better go," Wren said. "You shouldn't be late. This is one of those times you don't want to make an entrance."

Graham was seated in an auditorium with a few hundred people. It was better than he could have hoped. In a crowd this size, he wasn't going to stand out. He had been attending these speeches for three days, and no one had paid any attention to him. He hoped it was almost over. This was really boring. If he wasn't here for a reason, he would have left already.

Why did people want to work here so badly that they would come to a three-day conference? He hoped his friends were staying in the hotel room. Everyone at the conference was staying at the Gorbin Building, as it had suites on the top floors. They were comfortable rooms, but he was ready to leave.

After the third speaker, a tall, muscular man in a suit came to the microphone. Everyone clapped, and a few whistled. He ran a hand over his thick black curls and smiled.

"Yes, thank you, thank you," the man said. "If there is anyone here who doesn't know me, I am Gorbin Daniels, owner of Gorbin Corporations." More clapping followed. "Since you are all here, you obviously have good taste." Gorbin slyly grinned. "Unfortunately, we only need to fill thirty positions, so most of you won't make the cut."

Graham closed his eyes and tried to remember being captured at Meegore. If he wasn't mistaken, that voice belonged to the man who was leading The Dark Cloud on the island and at the mall when they had been attacked. If that was true, then people were going back and forth between Earth and Basura.

Graham's musing caused him to miss the motivational part of the speech. He tuned back in as Gorbin continued. "We will now break into groups. If you are here for a level A position, go to the west foyer and meet with Eddie. If you are here for a level B position, meet in the west conference room with Fria, and if you are here for a level C position, go to the east conference room with Nancy."

"Fria," Graham whispered. She was taking the level B's. Graham was a level A. If he followed the level B's, he would probably stand out as he wasn't old enough to apply for those jobs. Everyone started separating and Graham decided to follow her and hope he could get a hair before someone led him over to the other level A's.

Graham made his way over to the west conference room. A woman with a clipboard was walking around, checking off names. She was about forty, with long straight blond hair hanging to her waist. Her lipstick was a few shades too dark, and she was smiling so hard Graham worried it would break her face. She reminded him of a Barbie Kaylee used to play with.

"Welcome, welcome!" she kept saying, as she checked off the names. Graham tried to get behind her, but she was keeping her back to the wall. "I believe you are in the wrong group, sweetheart," she said when she looked at Graham.

"No, I'm in the right place," he said, hoping to see a blond hair on her shoulder.

She flashed a white, toothy smile. "You are a computer programmer?"

"Yes," Graham said, trying not to twitch. The other people in the room all laughed.

"I think you need to leave," she said, still smiling.

"Um ... can you tell me what that is?" Graham asked, pointing behind her. She turned around to see where he was pointing and he scanned her back. No loose hair.

"I don't know what you are talking about," she said, turning back to him. "Let me lead you to your group. I'll only be a minute," she told the others. Her high heels clicked against the floor as he followed her out of the room. In the hallway, her smile disappeared. "I don't have time for little boys who think they are funny. Go down that hallway and you will come to your group."

"Wait ..." Graham stammered. "I need ... a ... you have something in your hair." he reached out and yanked out a few pieces of her hair and he turned and ran.

"How dare you!" she screeched, chasing after him. He sighed and picked up speed. She was really going to chase him? In those heels? The hallway seemed to go on forever. He plowed into the door and raced outside.

When he reached the sidewalk, he stared back at the large brown building with shimmering windows. He put the hair into his pocket and smiled. He'd pulled out more than he intended.

Graham jumped when someone tapped his shoulder. He turned around to see Fria. She was smiling a victorious smile. "Look," she said, with a dangerous glint in her eye. "I'm tired of all of you kids thinking you can take things from me and sell them on eBay or something. Famous people deserve the same respect as anyone else!"

"Famous?" Graham asked, stepping back.

"Please. Don't play stupid with me. I'm Gorbin's second in command. You've seen me on the news, I'm sure. I know that's why you took my hair. You think you can earn some fast money?"

"Why would anyone pay for a few strands of hair?"

"Just give it back and I won't call security."

"Fine," Graham said, reaching into his pocket and pulling out some ailam powder. He blew it into her face and caught her as she collapsed. "Great, now what?" He looked around to see if anyone was watching. There had to be security cameras at a place like this.

A car pulled up, and the window rolled down. A man with a baseball cap, a red beard, and sunglasses peered out at Graham.

"What's wrong with your friend?" the man asked.

"Oh, uh ... too much ... a ... alcohol," Graham stammered as Fria snored.

"I'm an Uber driver. Do you need a lift?"

"Um ... sure," Graham said, not sure he had any other options. The man hopped out of the car and opened the back door. Graham carefully put Fria inside and climbed into the passenger seat.

"Where to?" the man asked, buckling his seatbelt. Graham handed him a piece of paper with the hotel address. "I never understand these people who drink like that," the man said, pulling away from the curb.

"Yeah, I don't either," Graham said, wiping sweat from his forehead. This was not going as planned.

"Are you from California?"

"Just visiting."

"How long are you staying?"

"Not much longer," he said, not wanting to tell too much to a stranger.

"You should probably go home soon," the man said, nodding. "It's a school day, isn't it? A vacation isn't worth missing too much school. If I were you, I would go home right away."

Graham raised his eyebrow and studied the man, but he couldn't read his expression with his sunglasses on. "I'm kind of homeschooling at the moment."

"So, who is that back there?"

"Step-aunt," Graham said, glancing back at the snoring blond woman.

"You should probably get her home as soon as possible."

"I plan to."

"Gorbin doesn't take lightly to people coming onto his property without permission. If he sees any video surveillance of your drunk aunt, he won't be happy. You need to make sure Gorbin doesn't come for her."

He pulled up to the hotel and put the car into park. Graham handed him some money and jumped out of the car. Something wasn't right, but he didn't know what it was. The man bounded from the car and helped get Fria out of the back. Graham picked her up and held her like a baby, and the man followed him into the hotel.

"I'm going to tell the staff what's going on so they don't bother you," he said, walking up to the front desk. Graham didn't wait around. He hurried to the elevator and rushed in. What a strange man.

He pounded on the hotel door, and Ming Li opened it. "Oh, great," she said, rolling her eyes. He came in and dumped Fria on the bed. "I hope that's Fria."

"It is."

"And you carried her in here like that?"

"Everyone thinks she's drunk."

"We wanted her hair, not her."

"I know. She chased me and I panicked. People are going to know where we are. We need to leave now. Where's everyone else?"

"In there," she pointed to the sitting room. "I can't believe you left me here with those people for three days. You owe me big time."

Graham entered the sitting room where Wren, Tal, and Sen were sitting on the couch watching a cartoon.

"Hey guys," Graham said.

Wren looked up with bloodshot eyes, and then back to the T.V. Tal and Sen didn't even acknowledge him.

"I accidentally captured Fria. We need to go home before Gorbin finds us."

"Not yet!" Tal said, still staring at the television. "I have to find out if the princess is going to forgive Estaban!"

"Oh boy," Graham muttered. "How long have you been watching this?"

"For almost three days!" Ming Li said, stomping into the room. "They won't sleep or anything. I tried to show them some other shows, but they kept getting freaked out. Even this cartoon made Tal scream once and Wren cry THREE times! I never would have turned this on if I would have known this would happen."

Graham walked over to the T.V. and turned it off.

"Hey!" Wren and Tal said together. Sen shook his head and seemed to be coming out of a daze.

"Guys, this is ridiculous. Did you hear what I said? I captured Fria."

Wren frowned, "Fria? You captured her?"

"Yes. You guys look like zombies. We need to take her back and have her locked up and you all need to sleep."

"I need to know what happens," Tal said, rubbing his tired eyes. Graham sighed as he noticed the dark circles.

"She forgives him. They catch the villains, and they all live happily ever after," Ming Li said, crossing her arms.

"Wow," Tal said, shaking his head. "That was intense. I feel kinda gross."

"You're still wearing the same clothes I saw you in three days ago."

"Wait, what's happening again?" Wren asked, standing up. "Whoa," she said, sitting back down, holding her head.

"They've been like this the whole time," Ming Li complained. "I fed and watered them every now and then."

"We have to go," Graham told them. "I have Fria here, and I'm sure Gorbin is going to come after us."

"Gorbin?" Tal asked, suddenly alert.

"Yes. He probably has security that will be able to track us. We need to go. Now. Wren, can you make a portal?"

"Sure," she stood again and put her thumbs and pointer fingers together and pulled them apart. Nothing happened. "I think I'm too tired," she apologized.

Graham ran his hands over his head and sighed. "Okay. You three all take a nap and I'll keep Fria sleeping. If anyone comes to the door, I'll wake you up and hopefully you'll be able to open a portal."

"Now go to bed!" Ming Li commanded, pointing to the bedroom. Sen was already asleep sitting up on the couch. "I can't believe you guys are making me be the responsible one."

Wren pried her eyes open and squinted at the bright window. How long had she been asleep? She couldn't believe she'd been so mesmerized by that magical screen. It was one of the most exciting things she had ever seen. She shouldn't have watched so much. Now she had a dull headache.

"It's about time you woke up," Ming Li said, entering the room. "You slept longer than Tal and Sen. We need you to make a portal so we can deliver Fria to Brake."

"Take Fria?" she frowned.

"Yes. Don't you remember? Graham captured Fria," she said, pointing at a sleeping blond woman on the bed next to hers.

"Right," she said, glaring at the woman. "I hope they put her in jail forever."

"So, are we going?" Tal asked, coming into the room with Graham and Sen right behind him.

"Are you still here?" called a gruff voice, followed by pounding on the hotel room door. They all glanced nervously at each other.

Graham peeked out the little hole in the door and sighed. "It's the guy who drove me here."

"I'll open a portal," Wren said, as the door clicked and a man with a long red beard and sunglasses came in.

He held up a key and shook his head. "I thought I told you to go home," he said, glaring at Graham.

"We are," Graham said.

"I left you here hours ago. You should already be gone. People are out searching for her," he said, pointing at Fria. "It's only a matter of time before they track you here."

"Wait …" Tal said, studying the man, "Are you Professor Dovin?"

The man sighed and took off his sunglasses and hat and threw them onto the bed.

"I can't believe I didn't recognize you," Graham said, taking a step back.

"Oh relax," Dovin said, pulling off the fake red beard and revealing Professor Dovin's smooth face. "You have all become a permanent pain in my side. Take that woman and get out of here."

"Why would you want us to take her?" Wren asked. She couldn't imagine handing over one of her friends, and he didn't seem to care.

"She's almost as big of a pain as the rest of you are. I wasn't ready to get rid of her yet, but you already set it in motion. I can't even believe the problems this is going to cause me."

"We should take you too," Ming Li said, with her hands on her hips. "It wouldn't be hard."

Dovin's mouth curled upwards into an amused smile. "As much fun as I would have proving you wrong, there isn't time. Wren, do your thing and open a portal. I don't care where it's to, but leave."

"Let's do it," Sen said, staring blankly at Dovin. "We were about to go anyway, right?"

Wren glanced at Graham and Tal and they both nodded.

"Great, thank you," Professor Dovin said, grabbing his glasses and fake beard. He put his hand on the doorknob and turned his head. "Please don't come back. I don't have time to keep cleaning up after the five of you."

"What are you talking about?" Tal asked.

"Another time, perhaps," he said, flashing his white teeth. "Oh, and congratulations, Sen. I always knew you would do big things."

Wren looked at Sen, but his face was like a blank slate. Dovin opened the door and left.

"So, portal?" Sen asked, glancing at Wren.

"I guess," she said, easily opening a portal into the jungle. Sen and Graham lifted Fria together and carried her through, followed by Ming Li. Tal grabbed a bag full of clothes and walked towards the portal. Right as he was about to enter, the silver portal shook.

"Whoa, what's it doing?" he asked.

"I don't know," Wren said, biting her lip and focusing on the portal. The portal sputtered and closed. "Oh, no," she said, trying again. Nothing.

"Are we stuck here?"

"I hope not."

"What's that noise?" Tal asked, going to the window. Wren followed and peered down. There were a bunch of the things Graham called cars around the hotel. They all had flashing red and blue lights and were putting off a loud noise.

"What do you suppose that means?" Wren asked.

"I don't know," he said, "But I have a feeling we're about to find out."

Graham watched as Brake, Zera, and Hamble studied Fria. She hadn't moved since they dropped her on the dirty jungle floor. Brake rubbed his chin, and Zera stood straight with an unreadable expression.

Hamble shook his head. "We searched for her for all these years, and it only took them a few days to catch her."

"Yes," Zera muttered, "But they were not supposed to capture her. That was dangerous."

"It wasn't really the plan," Graham explained, kicking at the dirt. "It's also not the biggest problem."

"Oh?" she asked, glancing to where Ming Li was sitting on a rock, sniffling.

"The portal shut and Wren and Tal never came through," Graham said. "We thought they might have closed it to grab something, but it's been ten minutes."

Zera let out a slow breath. "We should not be letting you do these things. It is too dangerous."

"What are we going to do?" Ming Li asked, holding back tears.

"There isn't anything we can do," Brake said, his eyebrows knitted together. "We tried to get to Earth for years and nothing ever came of it."

Graham shook his head. "Gorbin is coming and going, so there has to be a way. He was at Meegore. I recognized his voice."

"We don't know how he's doing it."

"We can't leave them. They must be in trouble."

"They're going to have to save themselves," Sen said, looking at Ming Li. "Wren is the only one who can open a portal. My father tried to get back to Earth for a long time and didn't have any luck until we met Wren."

"But there has to be a way!" Graham said, louder than was probably necessary. "They don't know anything about Earth. It will be a disaster!"

"Dryson ... Graham," Zera said, putting a hand on his arm, "Your father and I spent years trying to return to Earth. It is not a simple thing. Brake and Hamble can take Fria to the prison and the rest of us will wait here and hope

Wren and Tal appear. If they do not, we can try to come up with a solution. Alright?"

Graham looked into his mom's pleading eyes and nodded. "You can call me Dryson. A lot of guys on the basketball team called me Dryson, so it won't confuse me or anything." Zera's eyes sparkled, and she pulled him in for a quick hug.

"Professor Dovin was there too," Ming Li said. "So there must be a way to get there."

Brake narrowed his eyes. "What was he doing there?"

"He showed up when I hit Fria with the ailam powder. I didn't recognize him because he had a disguise. He helped me move Fria and came back later to tell us to take her and get out of there. He told us Gorbin was coming."

"Why would he help you take one of his own?" Hamble asked, his eyebrows coming together. "That makes little sense. There is more to Dovin than we know."

"He said he was planning on getting rid of her anyway or something like that," Ming Li said.

Graham studied Sen. "How do you know Dovin? Kids from Boztoll don't go to school in Akkron."

"He used to come to Boztoll and teach some of us kids," Sen said, shrugging. "We didn't like him because he only talked to the kids that were magic. My mom made me go study with him. Of the six of us brothers, only two of us have magic. He's the one who taught me how to warm things up and teleport around Basura."

"Did he ever say *why* he was teaching you?" Brake asked.

"He said it was important for those of us with magic to learn how to use it properly. He said there needed to be powerful people in Boztoll."

"Interesting," Brake muttered.

Fria groaned from the ground, and Graham jumped. He had almost forgotten she was there. Hamble took out a spray bottle and squirted her in the face. She flinched and wiped her eyes.

"Should I get some ailam powder?" Graham asked, reaching into his pocket.

"Nah," Hamble said, "I sprayed her with rednax venom."

"Stupid kids, taking my hair," she mumbled. She sat up, holding her head, and glared up at Graham. "You! How dare you? What did you do to me? I was only going to have you arrested, but now it will be much worse!"

"He is the least of your problems," Zera said, gritting her teeth.

Fria's eyes opened wide, and she jumped to her feet. "Zera!" she exclaimed, stumbling on her heels. She scanned the group and her face lost all color.

"It's been a while," Brake said, shaking his head. "And in many ways, not long enough."

"You sent that boy to capture me?" she squealed.

"I needed a piece of your hair," Graham said. "You wouldn't be here if you hadn't chased me."

"Why did you want my hair?" she asked, pulling her hair over her shoulder and stroking it. "Are you planning on putting a curse on me?"

"Oh please," Brake laughed. "All we are planning for you is prison."

"I won't go back with you," she said, folding her arms and tilting her head. "I'm an important person here."

"Here?" Hamble laughed, glancing around at the trees. "In the jungle?"

"No, in California," she scoffed.

"Well, it's good we aren't in California, isn't it?"

"We better not be in Akkron. I never liked you, Hamble," she said with a hostile glare.

Hamble smiled. "I know, and I'm sad about it."

"Why is my face sticky?"

"Rednax," he said, his smile growing. If Fria's face had been pale before, it was nothing compared to what it was now.

"No," she hissed. "No, no, no!" She turned and ran.

"I've got her," Hamble said, with an amused gleam in his eye. He sprinted after her. Graham was impressed. He never would have taken Hamble as a runner. It didn't take him long to catch up to her. She was going to regret wearing those heels today.

"It does feel like the old days," Brake said, rubbing the stubble on his chin as he watched Hamble wrestle Fria to the ground. "I hope it all ends better this time."

"I can't do any magic," Tal whispered as he and Wren hurried down the hotel hall.

"It must be Professor Dovin," Wren said, looking sideways at Tal. "He's blocked my magic before."

"But he was trying to get us to leave," Tal said. "Why tell us to leave and then block us?"

"I don't know. Maybe you should leave the clothes?" Wren suggested, pointing at Tal's bag full of clothing.

"No way. I love the clothes they have here."

"Where do we go?" Wren asked as they came to the stairs.

"I don't know. Let's focus on getting out of here."

They rushed down the stairs and into the lobby. There were people dressed in blue uniforms standing in front of the doors. Everyone in the lobby was staring around, looking uncertain.

One man stepped forward and held something out. "I am Officer Lopez," he said. "We are searching for a missing woman. No one is to leave this hotel until we have searched it. If you have any information, please come forward."

More people in blue began moving around the building. Wren felt a knot in her stomach. They needed to leave now.

"The doors are all blocked," Tal said, scanning the lobby. "Look at that guy over there."

Wren peeked to where he was pointing. A tall, muscular man was walking around acting bossy.

"Graham said Gorbin is one of The Dark Cloud members that captured them at Meegore and that fought us last time we were here. That guy looks like the description he gave us."

"Do you think he would recognize us?"

"It's possible. He's seen us in clothing from Earth and from Basura."

"If we go hide in a corner, that might draw more attention to us. What do we do?"

"I don't know," Tal said, putting his bag down. "I would say we fight without magic, but there are way too many of those people looking around. They must be guards or soldiers or something. We should wait and hope he doesn't spot us."

"We probably shouldn't be together. It might be harder for him to recognize us if we separate."

"Good idea. Why don't you go sit over there and read one of those books on the table and I'll stay over here?"

Wren nodded and walked over to an empty seat and sank down into it. She picked up a thin book and tried to read it. It was about fun things to do in the area. Most of it made little sense to her. You could go to a place and see princesses and have a magic adventure? It took her ten minutes to read it, and then she started over.

"Excuse me," a man said, standing next to her chair. She glanced up and saw the man who they suspected to be Gorbin peering down at her. "Have you seen this person?" He held an extremely realistic picture of Graham out to her.

"No," she said, going back to her book. That voice was definitely familiar. It must be Gorbin.

"Do I know you?"

"No."

"Really? You look very familiar. Something about your hair ... Yes, I see it. Where are your friends?"

"It's always about the hair, isn't it?" Tal said, coming up to them. Wren stood up, unsure what he might do.

"Where are the other two?" Gorbin said, flashing a triumphant smile.

"Three."

"What?"

"We have three friends now. We're expanding our horizons."

"Where are they?"

Tal smirked. "Gone."

"Where is Fria? You can tell me now and I'll be lenient with you."

"You aren't much of a threat to us right now."

Wren's stomach sank. What was Tal doing? Without magic, they would never get out of this.

Gorbin threw his head back and laughed. "I have the entire police force here at my disposal."

"Yeah, well, it turns out that even if our magic doesn't work here, ailam powder does."

"What?" Gorbin asked, glancing around. The lobby was full of sleeping people.

"I've learned to keep that stuff in vast supplies," Tal said, blowing dust into Gorbin's face. "Run!" he called to Wren. She raced off without looking back to see if Gorbin was asleep. They quickly hopped over the sleeping people in front of the doors and dashed out. They sprinted down the sidewalk, swerving around all the people.

"You still have your bag?" Wren asked, watching Tal run with the oversized bag.

"Hey, I paid a lot for all this stuff. Whoa, what is that?" he said, stopping. He pointed to an animal that resembled a horse, but with black and white stripes. Wren peeked behind them to see if anyone was coming.

"It's ten dollars to get a picture with the zebra," a man in a black and white shirt told them.

"This is a zebra?" Tal said, rubbing its head. "Is it a type of horse?"

"Really man?" the guy said. "You expect me to believe you don't know what a zebra is?"

Tal raised one eyebrow and smiled. "Whatever it is, she doesn't like you."

The man narrowed his eyes. "Get outta here."

"Let me buy her."

The man glared at Tal. "She's not for sale, now get."

"Are you sure?" Tal asked, holding out a handful of money.

The man's eyes widened, and he snatched the money out of Tal's hand. "She's yours."

"Let's get on," Tal said to Wren.

"You can't ride her," the man said. "She won't ever go for that."

"She said we can," Tal said, winking at the man and mounting the zebra.

"She's going to freak out and throw you," the man said, stuffing the money into his pocket. "She's big for a zebra, but they aren't meant to carry people. You could hurt her."

"Fine," Tal said, hopping off. He led the animal down the road and it trotted carelessly beside them. "Can you open a portal now?" he asked Wren.

"In front of all these people?"

"They have striped horses. What's weirder than that?"

"What is wrong with you?" Professor Dovin's voice called from the sidewalk. "How many chances do you need? GO HOME!"

Wren pulled her hands apart and opened a silvery portal to the side of them. Everyone on the streets turned to stare. Tal whispered something to the zebra, and it plunged into the portal with them.

# CHAPTER 11

"Wren!" Ming Li exclaimed, racing to where Wren and Tal had appeared with a zebra. Graham shook his head. Where did they get a zebra? Ming Li gave Wren a hug.

"What happened to you guys?" Graham asked. "Why do you have a zebra?"

"The portal closed and we couldn't do any magic," Wren told them. "We had to escape. Ailam powder is turning out to be a wonderful thing. I wonder why the adults didn't use it when they were fighting Gorbin."

Tal rubbed the zebra's nose. "We saw a man with this weird horse. She was miserable, so I bought her. I'm thinking rules about what is and isn't for sale aren't true on Earth. It's like how the person at the hotel told us we were too young to rent a room, we show a little money and bam. We have a room."

"You can't bribe everyone on Earth," Graham said, standing away from the animal. "And that's a zebra, not a horse."

"It's beautiful," Wren said, petting the beast. Graham shook his head. Zebras were cool, but he wasn't going to touch it. It had big teeth.

"It's pretty cool," Sen said, reaching out to pet it. "Where are you going to keep it, though?"

"I haven't thought that far ahead," Tal admitted. "Where's Fria? You didn't let her get away, did you?"

"Of course not," Ming Li said, rolling her eyes. "Some of the adults are taking her to the prison. Speaking of prison, shouldn't we be getting more praise or something? We've captured like four people. Five if you count Tram, although I guess that wasn't really us. He's not Dark Cloud either."

"Most of the adults are counting it all as dumb luck," Graham said. "With Karlof, it definitely was. We're probably lucky to still be alive."

Wren frowned, and Graham sighed. He shouldn't have brought up Karlof. Wren was still upset about her uncle's betrayal.

"Ouch," Tal muttered, as the zebra chewed on his hair. He moved away from her mouth. "We have Fria's hair. That was our quest, so I say this was a success. Capturing a Dark Cloud member is a bonus. Even Austra has to admit we're making progress."

"Progress, but you all sound sloppy," a shrill voice said, causing Graham to jump. He looked at Padmire.

"I thought you were going back to the cave?" Ming Li said.

"I was, but I thought about how far it was, and how useless my feet are on land, and I felt tired."

"You can stay here for a while," Graham offered. "You'll have to go into the meetinghouse, though."

"No, there is no time to waste sitting around in a meetinghouse. You all need to figure out what's going wrong with your weather and speed things up. If you don't fix it soon, how are people going to grow anything? Sure, you can live off of the other continent for a time, but that's not a suitable solution."

"We're making progress," Ming Li said, with her hands on her hips. "We finished the first step. You still have the hair, right, Graham?"

"Yes."

Padmire narrowed his eyes. "You over complicate things for mediocre results."

"How so?" Tal asked.

"You needed a hair from a traitor. Didn't you already have at least one traitor in your prison?"

They all glanced at each other. He was right. They had Karlof locked up. Getting a hair from him would have been simple.

"No, you aren't going to make us feel stupid," Tal said. "If we hadn't gone to Earth, we wouldn't have caught Fria."

"That was a pure accident from the way you told it," Padmire said, scratching his head as he glared at Tal. "And those britches you are wearing? I suppose you brought them from Earth. You realize they look stupid, right?”

“Hey Wren, can’t you make a portal for Padmire so he can go back home?” Tal asked.

She nodded. “If he wants me to.”

“No,” he said, shaking his finger at Tal. “I think I will stay around you people and make sure you don’t become a bunch of bungles.”

Graham wasn’t sure what to think. Having Padmire around all the time might get annoying. Still, he probably knew a lot of stuff that might prove useful.

“Now, what’s the next step in creating a silver eclipse?” Padmire asked.

“We need time to rest,” Tal said, as Sen pulled a folded paper out of his pocket.

“Rest is for losers,” Ming Li said, glancing at the paper Sen was holding. “It isn’t like we were doing much. You’re just worn out from watching too much T.V.”

Sen frowned at the paper. “I had to abbreviate it so it would fit on my arm. It says something about a large hedge, no deals, golden tear.”

“I don’t see any way to figure that out,” Graham said, running his hands over his face. Why did it have to be so complicated?

“I searched through a ton of books and didn’t find any-thing,” Sen admitted. “We could move on to the next one, which says something about water from a well of truth.”

“A golden tear. It sounds like the goblin maze,” Brog said from behind them. Graham jumped for a second time.

"Dang it, Brog," Ming Li muttered. "For someone so big, you can sure sneak up on people."

"Sorry," the giant said, with a small twinkle in his eyes. "I thought you saw me."

"What's the goblin maze?" Tal asked. "I feel like I might have heard of it before."

"It is a legend," Brog told them, as he kneeled in front of the zebra. "I have seen nothing like this before. What a wonderful creature." The zebra nuzzled its head into Brog.

"What's the legend?" Ming Li asked. "Brog? Earth to Brog?"

Graham grinned. "Shouldn't it be Basura to Brog?"

"Yes, the legend," Brog said, still admiring the zebra. "It was said there was a maze hidden in a valley between the goblin mountains. It was supposed to be beautiful, filled with lovely plants. There were creatures living inside it that would be helpful or misleading. It was better to ignore all advice and solve the maze on your own. Something of value was supposed to be in the center."

"A golden tear, perhaps?" Tal asked, tapping his fingers against his leg.

"There is mention of a golden tear in the maze, so it is likely."

"What are the goblins like?" Ming Li asked.

"They are tricky and untrustworthy mostly," Brog said. "They have strange magic that they occasionally sell."

"Like the portal dust Brake used," Graham said. He thought about the picture in his history book and shuddered. He hoped they didn't look like that artist's rendition.

"Please tell me we can go into a portal or Sen can teleport us there," Ming Li said. "I'm not up for a *Lord of the Rings* trek."

"A what?" Wren asked, furrowing her brow.

"It's not important."

"You can teleport to the goblin mountains," Brog said. "Probably not right to the maze, as we do not know its location. We must discuss this with the others. I would like their permission to accompany you there."

The corner of Tal's mouth twitched. "Brog? Can you say, 'she's,' or 'I've?' I'm just curious."

Brog's eyebrows knit together. "Of course I can. She is. I have."

"That's what I thought. I've always wanted to ask Zera the same thing, but she's kind of intimidating."

"Wasting more time," Padmire huffed.

"What is that?" Brog asked, pointing at Padmire.

"What?" he squealed. "What do you mean, what?"

"My apologies. I did not mean to offend."

"I am Padmire. I am a bungle."

"A bungle? Like the ones that were made to guard the cave at Meegore?"

Padmire puffed out his chest. "Exactly. It's good to see that someone knows about me."

"I did not know that bungles existed anymore, but I am happy to meet you. I am Brog, from Meegore." He bent down and offered his hand. Padmire shook his large finger.

"No more wasting time," Padmire said. "You all have a lot to do. I will also accompany you to the goblin mountains. Let's meet out here tomorrow after breakfast and we will begin our journey."

Ming Li crossed her arms. "I'm okay with that, but you know you aren't in charge, right?"

Padmire smiled, showing all of his pointy teeth. "Of course I'm not."

"That smile is not reassuring."

"Ben, leave Padmire alone!" Wren said, for the third time in ten minutes. Padmire was cowering against a tree root while Ben licked the back of his head. She scooped the little dragon up and put him on her shoulder. "Sorry Padmire."

"Dragons are disgusting creatures," Padmire said as he wiped the slobber off his head.

"He must like you," Wren assured him. "He only licks people he likes."

"Well, you better keep him away from me," the bungle said, glaring at Ben. "I don't see why *he* is coming with us."

"The adults only agreed to let us go if we take Brog, Austra, and Ben. Ben is sweet, but if he thinks someone's trying to hurt me, he can bite really hard."

"Where are the others?" Padmire asked, looking at the sun. "We should have left a half hour ago."

A door appeared out of nowhere and Tal, Graham, Ming Li, Sen, Austra and Brog came out. Everyone was wearing a dark cloak and had a small pack over their shoulders. They remembered to prepare this time and bring an extra change of clothing and food. Wren wondered if it was hard for Brog to find a cloak that big.

"Why are you late?" Padmire demanded.

Austra rolled her eyes. "We had to convince Talon to wear practical britches and not silly pajamas."

"I don't see how it would even affect you," Tal said, crossing his arms.

"I can't be seen with someone wearing something so ridiculous."

"Alright, let's get out of here," Padmire said.

"Everyone make sure you're touching my cape," Sen said, as they all gathered around him. Wren made sure Ben was secure as she grabbed a corner of his cape. "Everyone ready?"

Everything shifted. "That shocks me every time," Graham said, as they appeared at the base of a gigantic mountain.

The ground was covered with a light sprinkling of snow. Wren was glad she wore her knee-high boots. "That's an enormous mountain," she said, gazing up.

Ming Li shook her head. "I knew it. *Lord of the Rings.*"

"It's going to take us forever to get up there," Tal said.

"I need to be carried," Padmire said. "My feet will freeze." Brog picked him up and placed him on his broad shoulder. Padmire sat and grabbed hold of Brog's brown braid and wrapped it around himself.

"You are all ridiculous," Austra said, clapping her hands three times. "It's a good thing I've come with you or, I am assuming, you wouldn't live through this." She stomped her feet four times and let out a shrill whistle.

"What are you doing?" Ming Li asked, covering her ears.

Austra cocked her head and stared at Ming Li. "I'm saving us days of climbing that mountain."

A bright white line appeared six feet in the air and slowly grew bigger, zigzagging to the ground. Everyone stepped back and watched in surprise as the zigzag split in half and pulled apart. Out of it stepped a small, wiry, gray goblin. He couldn't be taller than four feet. He was bald and had a cloth around his waist. It was the only thing he was wearing. He didn't even have boots. His eyes appeared too big for his face, and he glared at each of them. Wren tried not to flinch as he studied her.

"I am Vork," he said, glaring at Austra. "Why have you summoned me?"

"Hello Vork," Austra said, bowing gracefully. "I am Austra."

"Yes." he said, squinting up at her. "I remember you. Not as harmless as you appear, are you?"

Wren tried to remember something someone had said about Austra and goblins. If she remembered correctly, Austra had been in a fight with a few of them. A fight she had won.

"We haven't come to cause trouble. We have come for your help."

"Why would we help you? I remember your group. Always searching for easy magic. And now you come here with children ... among other things," he said, as he sneered at Brog and Padmire.

"I see you are suffering from the weather, the same as we are," Austra said, gesturing at the snow. "We are seeking a solution to this problem."

"And what solution do you seek here?"

"We are on a quest to create a silver eclipse."

"Silver eclipse ..." Vork said, rubbing his bumpy chin. "I have not heard of that in a long time. Many have tried over the years and failed. What makes you think you can do this?"

"They have me," Padmire said, still wrapped in Brog's hair. "I have given them the directions to make the eclipse."

"And where did you get those directions?"

"The cave at Meegore."

"I see," he said, rubbing his hands together. "I still don't understand why you are here."

"We need to go to the maze," Austra said.

Vork laughed. "The maze is only a legend. You've come here for no reason."

"If we don't go into the maze, we can't make a silver eclipse," Wren said. "The weather will get worse and worse and everything will die."

Vork studied her with his enormous eyes, and she could almost see the wheels turning in his head. "Fine," he finally said, throwing up his hands. "But we will *not* be responsible for anything that might, and probably will, happen to you."

"That's fine," Wren said, trying to keep Ben from jumping onto the ground.

The goblin waved his hand, and the ground shook. Everything went black for a moment and when they could see again, they were in a grassy area in front of a fifteen foot hedge. There was no hint of winter in this place. Wren was surprised the goblin hadn't demanded anything. From what she had heard, goblins never did a favor without something in return. They must be just as scared about the weather as the rest of the continent.

"No one has been inside in two hundred years," the goblin said. "And no one has come out in even longer." Wren shivered. It seemed so green and peaceful. "I can only assume you need a golden tear. If you go in and succeed, you also give us a golden tear. That is my deal." There it was. He did want something.

"Why don't you get your own?" Graham asked.

"Nothing about the maze is clear. It has been so long since anyone has entered. We do not know all the dangers and we are not willing to go into danger."

"What if we can only get one?" Tal asked.

"Then you must fight to keep it," Vork said, baring his yellow teeth.

"Please," Austra snorted. "That shouldn't be too hard."

Vork glared at Austra and pulled his hands apart. The hedge opened. "There are many of you. The hedge does not like too many people together. It will separate you, so be prepared."

"Separate?" Wren frowned. "That doesn't sound good."

"Some of us should wait out here," Graham said. "Then if the people inside need help, they can signal the others."

"I hope you don't mean signaling with magic," Vork grinned.

Tal rolled his eyes. "Don't tell us. Magic doesn't work in the maze."

"Smart boy."

"If anyone stays behind, it should probably be Austra and I," Brog said. Austra started to protest, but Brog shook his head. "They are the ones chosen. They are the ones who need to go in." Austra pursed her lips and nodded.

"That still makes six going in," Ming Li said. "Seven if you count Ben. Is that any better?"

Vork shrugged.

"I will stay here as well," Padmire said. "I don't want to risk getting lost in there."

"If you don't come out in two hours, we will come in and find you," Austra said.

"Two hours?" Vork laughed. "They will not be out in two hours. You cannot see the maze. It is large. The fastest I ever saw it completed was in three days."

"Three days?" Ming Li and Graham exclaimed together.

"We definitely don't have enough food to last that long," Tal said, peering into his pack. "If we ration it, it might last for two."

Vork rolled his eyes. "You won't starve in the maze. Look at it. It's covered in berries. They are throughout the whole thing. There is also a small stream that flows through. It's only a trickle, but it will keep you alive."

Wren looked closer at the green foliage and noticed the small red berries. "Are you sure the berries are safe?"

"Of course they're safe. We goblins don't go inside the maze, but we eat the berries that grow on the outside. They're delicious."

"After three days of eating berries, I can only imagine the digestional issues," Tal said, cringing. "And I'm guessing there aren't any bathrooms."

"You all seem to think you will finish in three days," Vork laughed. "I said the fastest was out in three days. The longest was two months."

Graham groaned. "Two months? We don't have time for that."

"Two months," Vork said with a grin. "Unless you count the ones that never came out."

Tal gazed into the maze. "So, are we searching for a literal tear or something that looks like a tear?"

"How should I know?" Vork muttered. "I've never seen one."

"Then why do you want one?"

"Anything gold is worth having. Not worth risking *my* life, though."

"Can we hurry and do this before he totally freaks us out?" Ming Li asked. "I don't know about the rest of you, but the more he talks, the less I want to go in."

"What if it does take months?" Wren asked. "Brog and Austra can't sit out here forever."

"I can," Brog assured her. "I have been trained to stay in one place for a long time."

"I can as well," Austra said. "But try to be quick. Remember, this is only one step for the silver eclipse. There is still a lot to do, and the world is counting on you."

# CHAPTER 12

"So, which way do we go?" Tal asked as they all studied the three paths in front of them.

Graham glanced to the right and then to the left. It all looked the same. The ground was dirt, and the hedge was green. That was about all they could see, except the blue sky. This was going to be almost impossible if the entire thing looked like this.

Ming Li pointed forward. "Let's go that way. I mean, we haven't tried anything, so any choice is as good as another."

"We should go right," Graham said. "I remember hearing once that you could always get out of a maze if you always turn right."

"That sounds logical," Tal said. "Let's hope it isn't magic or anything."

"It probably is. If it wasn't, there would be snow. It's not even very cold in here," Graham said.

"Should we run or walk?" Sen asked. "I don't want to be here forever."

Wren's mouth turned down. "If we run, we might wear ourselves out and not get anywhere."

"Or we might get finished faster," Ming Li said.

"Walking would be better than standing here," Tal said. "Let's go." They turned right and walked at a fast pace. Ben didn't want to be carried, so he bounded ahead of them. "What if Ben doesn't go right?"

"It's possible he has some kind of animal sense," Wren said. "Maybe he can find the right way."

"I don't think so," Tal said. "He's just excited to be running around."

"Did he tell you that?"

"I can feel it."

They turned right again, and Ben ran faster.

"Ben! Come back!" Wren called as Ben disappeared around a corner.

"I'll stop him," Graham said, running to the spot Ben had disappeared. He turned the corner and saw the green dragon chasing a butterfly. Now what? Graham never touched animals. Why had he volunteered? Any of the others would have been a better choice.

"Come here Ben," he said. Ben turned and stared at him, and trotted over. "Good dragon. Now, go that way." He pointed behind him. Ben studied him and tilted his head. "Come on. Let's get Wren." Ben walked up to him and licked his boot. "Gross. Go that way." Graham turned around and pointed, but instead of a path, there was a dead end.

"Oh, no!" Wren said, running to the place Graham and Ben had disappeared. "Graham! Can you hear me?" Silence answered.

"We can try to push our way through," Ming Li said, pushing at the hedge. "It's so thick."

"This is terrible," Wren said. "We all need to stay close together. I've never seen anything grow that fast. Well, maybe when Tal grows something ..." She turned to look at Tal. He was gone. "Where's Tal and Sen?"

"What?" Ming Li asked, pulling her hands out of the hedge.

"They're gone!"

"Sen!" Ming Li called. "TAL! GRAHAM!" they listened. Nothing.

Wren bit her lip. "Great. I'm guessing there isn't any reason to search for them. It's probably like Meegore, and we'll all end up in the same place."

"You know Graham is with Ben."

"I know," Wren said, wringing her hands.

"He still doesn't dare to touch him. Do you think they'll stay together?"

"I sure hope so. Let's stick to Graham's plan and keep going right." Wren didn't want to think about Ben getting lost in the maze. They kept walking and turning right for what seemed like forever. Nothing ever seemed to be different. Wren wished they could use magic. Tal could probably part all the hedges and get them right through.

"What is that?" Ming Li asked, pointing over Wren's head. She glanced up. "Is that a fairy?"

"Wow," Wren said, gazing at the sparkling fairy. She flew down and smiled at them.

"Hello," Ming Li said. The fairy kept smiling. She was wearing a long pink dress and her hair fell in soft golden curls. On top of her head was a glittering tiara. "Who are you?" Ming Li asked. The fairy turned and flew ahead. She motioned for them to follow.

"Do we follow?" Wren asked, looking at Ming Li.

"I don't know. I guess."

"What if it's a trick?"

"I guess we'll find out."

"Alright," Wren sighed, focusing forward. "Wait ... wasn't it this way?" Wren pointed at the hedge in front of them. She turned around.

"Where did she go?"

"I don't know, and now this way is blocked."

"I guess that means we go this way then," Ming Li said, walking back in the direction they had come.

"I hope everyone else is doing better than we are. Let's make sure we stay close so we don't end up separated." They linked arms and started walking. They took a turn that led to a long pathway. "This is going to take a while."

"I'd rather make turns than walk straight. This seems to go on forever," Ming Li complained as they walked. After what seemed like an eternity, but was probably only fifteen minutes, there was a four-way split in the path.

"Let's go this way," Ming Li said, pointing to the right. Wren nodded, and they turned.

Wren saw something out of the corner of her eye. She spun around but there was nothing there. "I thought I saw something."

"Like what?"

"I don't know. A mouse?"

"Gross. There are probably tons of rodents in a place like this. Especially if people haven't been in for a long time."

"It seems well kept though. I wonder who keeps it trimmed and nice."

Ming Li shrugged. "It's magic, so it's likely it keeps itself nice."

"I guess," Wren said as the path curved.

"Is it time to eat yet? I have no idea how long we've been here."

"An hour? Possibly two."

"It feels longer."

The fairy flew in front of them and motioned them forward. Wren looked at Ming Li and shrugged. "Do we follow?"

"Sure."

The fairy was flying fast. They had to sprint to keep up.

"Look, there's an opening! Is that a way out?" Ming Li asked as the fairy flew over the hedge and disappeared.

Wren squinted at the opening. "I'm not sure." They walked over and peered out. "Oh no! We are back at the start."

"Is something wrong?" Austra asked, getting to her feet. Brog was sitting on the ground and Vork was standing with a mischievous glint in his eye.

"We were separated," Wren explained. "We saw a fairy, and we got turned around."

"Hee hee," Vork laughed. "Never pay attention to a fairy. They are only up to mischief."

"Let's only go left this time," Ming Li said. "And we won't let anything distract us."

"Be careful," Austra said. "I don't know what I would tell your father if you didn't come out."

"Come on," Ming Li said, linking arms with Wren. "Let's go fast and make sure we don't lose each other."

"I already feel like we've been walking forever," Wren said as they followed the walls and turned left. "I bet only turning one direction doesn't work if the maze can change."

"Any ideas?"

"No."

They walked in silence for a few minutes, automatically turning every time there was a left turn. Wren saw something ahead. It looked like an animal, but it was too far away to know what kind. It dashed away before she could make out what it was.

"Did you see that?" Ming Li asked, with wide eyes.

"Yeah," Wren said, turning to her friend. "Do we keep going?"

"Not that way," she said, pulling Wren to a stop.

"Great. Another hedge." How had it popped up so fast?

"I think we have to watch ahead. It only seems to change when we look away."

Wren sighed. "That's going to be bad if we have to sleep here. What will happen when we aren't watching anything?"

A massive lion leaped out in front of them. Its golden mane flowed around its majestic face and his amber eyes sparkled unnaturally. They both screamed as his roar shook the maze and filled Wren with a deep foreboding fear. His eyes pierced Wren's, and then he scampered away without another sound.

"This is so stupid!" Ming Li half yelled. "We are going to be lost here until we get eaten by a lion!"

Wren put a hand to her chest. Her heart was trying to bust through. "I wish we were with Tal. He might still be able to sense animals. Did you see those eyes? I've never seen anything like that. They didn't seem real."

"I wasn't focused on the eyes. Why is there a lion? Why are there fairies?" Ming Li muttered, running her hand through her long hair. "If no one has been in here for two hundred years, why are there still weird things in here?" She pulled open her pack and took out a cookie. She bit into it so hard Wren could hear her teeth grind together.

"Are you going to be okay?"

"Yes," she muttered. "Sorry, I'm not used to letting things like that scare me."

"Should we run?"

"Maybe," she said, stuffing the rest of the cookie into her mouth. "How about run one minute, walk one minute?"

"Okay, but we can't take our eyes off the path."

"Let's do it."

"Slow down!" Graham called to Ben. This was ridiculous. Ben could run forever. It was a good thing he became distracted so easily so Graham could keep up. Ben stopped to sniff something on the ground, giving Graham a moment to rest. He put his hands to his knees and breathed hard. They had been running off and on for almost two hours. He wasn't sure how much more he could take. He thought about going without Ben, but he knew how much Wren cared about the little devil.

Graham made sure he kept his gaze ahead. It hadn't taken him long to figure out that every time he looked away from the path, a hedge would grow up in front of him and he would have to backtrack. He wouldn't swear to it, but he thought he saw a lion running across the path a little earlier.

"How did I get stuck with you?" Graham asked as Ben sniffed around his feet. "We need a new strategy, because you're wearing me out." Ben cocked his head to the side and stared at Graham innocently. He blinked twice and then climbed up Graham's leg. "Hey! What do you think you're doing?" The dragon twisted around him on his way up and stopped at his shoulder.

"He likes you."

Graham spun around. A thin, brown-haired boy who appeared to be about ten was standing in the path behind him. He was wearing a scruffy brown tunic and brown boots. "Who are you?"

"Dragons are useful creatures," the boy said, ignoring his question. "They are loyal. Are you loyal?"

"Um ... I think so," Graham said, trying not to think about Ben's teeth so close to his face.

"You don't seem sure."

"I'm sure."

He tilted his head to the side and studied Graham. "If you were loyal, would you have lost your friends?"

"That had nothing to do with being loyal," Graham frowned. Who was this kid? "We were separated."

"But you wouldn't have been if you had stayed close."

"Where are you from? What are you doing here?"

"You should probably find your friends. Things are always better when you work as a team."

A shiver ran down his spine and goosebumps jumped full force onto his arms and legs. "Yeah, I'm gonna go." He had seen enough scary movies to know you should stay away from creepy little kids. The boy's expression hadn't changed once. There was no way that was normal. He was more stoic than Sen.

"Do you want to know where your friends are? I can help you."

Graham took a step away. "Oh?"

"If you give me your dragon, I will take you to them." He reached out towards Ben and the dragon growled.

"No." Wren would never forgive him if he traded her beloved pet for information, or anything else for that matter.

"But you fear him."

"I'm going to go," Graham said, turning and rushing away from the boy, or whatever he was. A normal kid couldn't be running around this maze for over two hundred years.

"Come back if you change your mind."

"No chance," he said as Ben nuzzled against his neck. "Yeah, yeah. This doesn't mean I like you. We can come to a truce, though, if you promise not to bite me. Or slobber on me." They turned a corner and kept walking. Having Ben on his shoulder might be a good thing. At least it meant he got to choose the direction and the pace, but for being a small dragon, he was still heavy.

He rounded the corner and almost bumped into the boy. "Ahhh!" he screamed, jumping back. Ben bared his teeth. "How did you get over here?" he asked, backing up.

"I thought you might want to trade something else. Your friends are all doing much better than you are. If you make a trade, you can be with them again. All you have done is run in circles. You are no closer now than you were two hours ago."

"Have you been following me?" This kid was creepy. He appeared fairly normal if you didn't count the blank expression on his face. That was what made him scary. Nothing seemed to be going on behind those eyes.

"I will trade you information about your friends for your cloak."

"No deals," he said, remembering that part of this quest had come with that warning.

The boy nodded and faded into smoke.

"You saw that, right?" Graham asked, focusing on Ben. Ben growled as he watched the smoke blow away. "We need to get through this fast. We need to find Wren. Can dragons follow scents?" He placed Ben on the ground. "Find Wren." Ben sniffed the ground and ran. "Great. Running again."

Graham followed the dragon. "Slow down, Ben! I'm wearing boots, not running shoes." Ben turned around and looked at him, but continued running. He seemed to slow down a little. Graham wasn't sure, but he thought they might have already passed this way. So much for strategy.

Wren stared up at the stars and tried not to feel discouraged. They were bright. It had been a long time since she had seen stars like this. There must be something magical here, keeping the weather nice and the sky clear. She hoped there weren't any bugs on the ground. The cloak under her cheek was soft, but still far from comfortable. Both she and Ming Li had rolled up their cloaks to use as pillows. It was warm enough without a blanket, but Wren wished she had one to keep any unwanted creatures off of her. Ming Li was lying next to her, the strap of her bag wrapped around both their legs as a makeshift tether to keep them from being separated during the night.

Ming Li yawned. "I shouldn't have worn new boots. I'm gonna have blisters."

"Me too," Wren said. "My legs are killing me. I thought it would be better to have a lower heel, but my muscles aren't used to them."

"Do you think the others are okay?"

"I hope so."

"Tal can probably keep a lion from eating him, but what about everyone else?"

Wren frowned in thought. "Can Tal still sense animals? Vork said we can't do magic here."

"I don't know."

"What other spine chilling things do you think are in here?"

Wren grinned. "Spine chilling?"

"Lions scare me."

Wren turned her head to look at her friend. "We've known each other a long time, but it's only been since Graham came that I've realized you are scared of things. You always act so brave and carefree."

Ming Li chuckled. "I completely believe in the 'fake it till you make it' mentality. I'm scared of way more things than I let on. If I pretend I'm not scared, then I'm not as scared. Does that make sense?"

"Sure. I hate my basement at home. There isn't a light on the staircase, so every time I walk up in the dark, I force myself to stay calm and walk steadily. Then I tell myself I'm not scared and there is absolutely nothing in the dark behind me waiting to grab my ankle before I get to the top."

"Does it work?"

Wren sighed. "No, not really. I probably don't seem scared, but on the inside, I feel pretty sure something is going to get me."

Ming Li chuckled again. "I kind of have that feeling right now."

"You can't be scared. If you're scared, then I feel more scared. Can we talk about something else?"

Wren felt Ming Li stiffen next to her, then sit up. "Did you hear that?"

"Ming Li!"

"No, I'm serious. Listen."

Wren strained her ears. A rustling on the hedge to the right of them made her heart stop, and all the hairs on her arms stood on end. She sat upright and clutched Ming Li's arm. "What is it?"

"I don't know," her friend said, holding on just as tightly to Wren.

They stood slowly and peered up at the hedge. The sound was up high, and they could see the leaves shaking in the moonlight. A dark shape leaped to the ground. Wren screamed and tried to run, but both of them ended up tripping on Ming Li's bag strap and falling to the ground in a heap. They squealed and tried to untangle their feet.

"Calm down," Sen's voice said, breaking them out of their panic.

"Sen?" Ming Li said, glaring up at the boy. "You scared us to death!"

He shrugged. "Sorry. I didn't know you were here."

"What were you doing?" she growled.

"Tal and I weren't getting anywhere, so I decided to try a different approach."

"You climbed that?" Wren asked, pointing up at the tall hedge. "It's huge!"

"Not that hard. It's pretty strong, so it doesn't bend much like I thought it might."

"Where's Tal?" Wren looked up, trying to see her friend.

"No idea. I told him to come, and he told me it was a waste of time and energy. I've been climbing hedges for over an hour. He should have followed, but he didn't."

Wren cupped her hands over her mouth and yelled, "TAL! TAL CAN YOU HEAR ME?"

No answer. "Should we try to climb to him?"

Sen shook his head. "From what I've seen today, he wouldn't be there. I left him a while ago and this place is strange."

"If we end up running around in this place for two months, I will not be happy," Ming Li said. "It's only been one day and I am ready to quit."

"No way," Sen said. "You don't strike me as a quitter."

"I'm not, but this is lame."

"So now Tal's all by himself. He's probably so scared," Wren said, scanning the top of the hedge.

"I'm sure he's fine," Sen said, waving a hand dismissively. "That guy never stops talking. I can't think when I'm with him because he never shuts up."

Ming Li grinned in the dark. "Everyone talks a lot compared to you. I must drive you crazy. I talk way more than Tal."

"You don't drive me crazy. At least you seem to care about what you're saying. He likes to hear himself talk. Like, does he really care what my favorite color is?"

Wren smiled. "He asked you what your favorite color is? While we're lost in the maze?"

"Yes. And my favorite animal, and anything else he could think of."

"He's probably just trying to get to know you better."

"Out of curiosity, what is your favorite color?" Ming Li asked.

"Yellow."

"Yellow?" Ming Li asked, scrunching up her nose. "I didn't know anyone's favorite color was yellow. I would have guessed blue." Sen shrugged.

"We should probably sleep," Wren said, rolling her messy cloak back into a pillow. Her heart was still beating fast, but she was trying to ignore it.

Ming Li reached for Wren's pack. "Now we need another pack to tie Sen to one of our legs."

Sen blinked. "Excuse me?"

"It's so we don't get separated while we're sleeping."

"Is that why you two were tripping all over each other?"

"Yes, but you scared us."

"It isn't safe. If something threatening comes, I want to be able to jump up and not trip."

Wren bit her lip. He had a point.

"Here," Sen said, spreading his cloak on the ground. "We can all make sure we're at least part way on the cloak, and that should keep us connected." Wren wasn't sure, but she thought he might be rolling his eyes. She wondered if he got scared of things. He didn't seem easily ruffled.

"Tomorrow, we solve this thing," Ming Li said, lying down. Wren hoped she was right.

Morning finally came. Graham pulled his cloak tighter around himself and forced his eyes to open. Ben was curled under his arm. That had been one of the longest nights Graham could remember. Part way through the night it had gotten chilly, making it hard to sleep. The dragon didn't seem to produce much heat, but it was warmer to have him close.

They seemed to have developed a new relationship in the last seventeen hours. Graham wasn't worried about getting bitten at all anymore. Slobber was a different story. Who would have thought that dragons would slobber? It didn't seem right. Ben yawned and stood up.

"You couldn't make a little fire?" Graham asked, as he sat up and brushed at his hair with his hand. His curls were starting to hang over his forehead. He usually kept it pretty short, but they had been so busy lately.

"It's about time you woke up."

Graham jumped to his feet and looked up to see Tal on top of the hedge.

"What are you doing up there?"

"I have no idea," he grumbled. "Actually, Sen climbed over the hedge yesterday and ditched me. I thought I would try it. You haven't seen him, have you?"

"No. Are you going to come down?"

"Probably," Tal said, glancing down at him.

"How are you sitting up there like that? You don't even seem like you're sinking."

"It's surprisingly solid. I mean, it's a plant, so it's kinda poky and stuff. I wonder if we could crawl on the top to the center."

"Are you only saying that because you're too scared to come down?"

"What? No. Actually, yes. Up seems a lot less freaky. There aren't really hand holds, so you have to take a handful of the hedge and hope it's strong. I wish we could use magic. That would make it so much easier." The hedge under Tal started sinking. "Whoa!" he said, grabbing hold. "Some magic must work," he said as it lowered him to the ground.

"Did you tell it to do that?" Graham asked as Tal jumped off and the hedge sprung back up.

"No, but my feelings are connected to the plants. It probably sensed I wanted to go down. I haven't been able to pick up any feelings here, though."

"So now Sen is out there somewhere by himself," Graham said as Ben galloped in circles around Tal. He bent over and picked him up.

"Traitor," Graham said to the dragon.

Tal laughed. "Have you and Ben been bonding?" He rubbed the smooth scales on Ben's head. "You know I'm always going to be the favorite, though."

"I don't know. He seems to like Brog."

"Yeah, all the animals like Brog."

"Is he going to take charge of your zebra?"

"I hope not, but it's hard to take care of animals when we have so much going on. You know about that, you let Brog take Walter."

Graham thought about his dog. "Yeah, Walter loves him. I've never been an animal person."

"That's an understatement."

"I like Walter, it's just, I don't know. He kind of got willed to me. Most animals make me nervous. Ben and I have come to an understanding though."

"Well, that's good. I'm sure Wren will be impressed."

Graham shook his head.

Tal looked around. "Well, if I had to lose Sen to hang with you, it's worth it. Have you ever tried to talk to that guy? He has no people skills at all. I tried the best I could to make conversation, but he would only give one-word answers and act annoyed."

"What did you ask him?"

"Everything I could think of. His favorite color is yellow. Whose favorite color is yellow?"

"What's wrong with yellow?"

"I don't know. It's the color of jaundice."

"So you just spouted off questions?"

"Pretty much. Silence makes me feel awkward. I figured I haven't tried hard enough to get to know him, but man, he does not make it easy. He did tell me about his name. So, his dad's name is Rosendo. He wanted to name Sen after him. Isn't that the weirdest thing you've ever heard? I would hate having the same name as my dad. That would be so confusing."

"People do it on Earth sometimes."

Tal rolled his eyes. "Well, I think it's stupid. It was smart to make it a law that nobody can have the same name."

"That seems hard. I bet some people go crazy trying to make up a name that doesn't already exist."

"You can use a name that exists. The last person who used it has to have been dead for at least fifty years. Anyway, since Sen's dad couldn't name him Rosendo, he named him Sendo."

"I guess that sounds similar."

"Oh, not you again," Tal said to someone behind Graham. Graham turned around to see the boy from the day before.

"I don't think he's real," Graham told him. "He turned into smoke or something yesterday."

"I see you still haven't found all of your friends," the boy said. "I can help."

The corners of Tal's mouth turned down. "Leave us alone."

"None of you are even close. Most of you are staying in the same wrong corner. At this rate, you will die here."

"This whole place looks the same," Tal complained. "I could easily believe I've been passing the same thing over and over."

"You are doing better than the friend who abandoned you. He kept jumping over the same hedge. He thought he was making progress."

"What are you?" Graham asked, as he fastened his cloak. He wanted to be ready to run.

"If you don't want to make any deals, I will tell you only one thing. Don't let the labyrinth know what you are planning to do." He turned around and walked away.

"Labyrinth?" Graham said, tapping his fingers against his pack.

"A labyrinth is like a maze," Tal explained.

"I know, but I thought a labyrinth only had one way out. And a minotaur at the center ... or possibly David Bowie."

"I understood nothing about that last sentence."

"I'm hoping there isn't a monster in here, or a dead singer my aunt used to have a crush on," Graham clarified.

"Sen said he saw a lion, but when I turned, there wasn't anything."

"I thought I saw a lion."

"Great. We should have brought weapons to make up for not having magic."

"If we aren't supposed to let the maze know what we are doing, let's put one of us in charge and the other person can follow."

Tal nodded. "You lead."

Graham took an apple out of his pack, slung the bag over his shoulder and walked.

"There's a nursery rhyme about a labyrinth," Tal said, as he followed.

"Yeah? How does it go?"

"I'm trying to remember. It's something like, through the labyrinth, pretty maids. Umm ... watch the path, or it might fade. Something about glowing eyes? Twinkling eyes? Follow the eyes? I can't remember. I've never been big on rhymes."

"Try to remember. It could be important. If it says something about a path fading, it could be about this place. The paths don't fade, but they sure disappear." He took a bite of his apple and tried to think and walk at the same time. Should he be using a strategy of walking aimlessly? He stepped over the small stream of water that ran through the maze. It crossed their path so many times he didn't even think about it anymore.

"What does Ben eat?" Tal asked, offering a grape to the dragon. He devoured it and gazed up at Tal expecting more.

"It looks like grapes. I heard Austra say something about hoping he was eating all the mice."

Graham stopped in his tracks. "Lion," he said, pointing ahead.

Tal squinted at the lion. "I'm trying to get a feel for him, but I'm getting nothing." The lion stared at them.

Graham cocked his head and studied the enormous beast. Its eyes were almost glowing. "Follow the eyes."

"I don't think that's a good idea."

"Your rhyme. You said to follow the sparkling eyes or something."

"I could be completely wrong. It could have said, don't follow them. I can't remember."

The lion turned and walked away. Graham clenched his jaw and took off after it. Tal didn't appear happy, but he followed.

"This could be the worst decision you ever make," Tal said, handing another grape to Ben.

Graham kept going forward. He hoped that wouldn't be the case. The lion ignored them as they followed him left and right for what felt like forever. Graham's toes were aching in his boots and sweating.

"Where do you think he's going?" Tal asked. "I don't know how long we can keep this up. He might be leading us away from where we want to go."

"Maybe, but he did have sparkling eyes," Graham reasoned, not slowing down. "I could see that even though we weren't close. I wish you could remember the poem."

"I've been trying. The more I think about it the more I mix it up. I wish he would stop for a breather."

"He probably doesn't need one. He's obviously magic if he's been here for hundreds of years."

The lion stopped, causing Tal and Graham to stop. He turned and watched them with his unnatural eyes. Graham sucked in a breath. He hoped it wasn't hungry.

"I wish I could sense him," Tal said from under his breath.

"Can lions climb?" Graham asked, stepping closer to the hedge.

"They can, but they aren't good at it. He doesn't look like he's going to come after us though. He seems calm." Tal took a step forward. The lion ignored them and walked away. "It looks like we're moving."

# Chapter 13

"These berries are so good," Ming Li said, tossing one in her mouth as they walked. "I could eat them until I burst."

"They are pretty good," Wren agreed. "Hey, Graham should put some in his medicines. They might be powerful enough to mask the taste."

"That's a good idea. We could give some to Brake and tell him to put them in his muffins. It would have to help."

"Why doesn't someone teach him to cook?" Sen asked.

Ming Li's lip curled up in a smirk. "Have you ever heard the expression, you can't teach an old dog new tricks?"

"No. But Brake isn't a dog, and he doesn't seem that old."

"I bet Zera did all the cooking, and when he was living by himself he didn't ever try hard. Wren, are you ever going to stop whistling that?"

"Sorry," she said. She hadn't even realized she was whistling.

"What is it anyway?"

"It's a children's song my dad used to sing."

"My mom used to sing it too," Sen said. "It's actually crossed my mind a few times since we've been here."

"Why?" Ming Li asked. "What does it say?"

Sen cleared his throat and sang,

Through the labyrinth, pretty maid,<br>
Never let the pathway fade,<br>
Follow the Beast with sparkling eyes,<br>
But do not let it hypnotize.<br>
When you make it to the end,<br>
Follow your heart and be a friend,<br>
Difficult choices will arise,<br>
Please choose wisely and win the prize.

"Wow, you can sing," Ming Li said, and Wren nodded in agreement. "If you were both thinking about that song, why didn't you bring it up? Now we know what to do."

"What are you talking about?" Wren asked. "It's only in my head because we're in a maze."

"I bet it's about this place. So now we need to find the freaky lion with the crazy eyes and follow it."

Sen nodded. "I knew I saw a lion."

"I don't see that following a lion is a great idea, but we aren't making any progress," Wren said, wringing her hands.

"I was more freaked out by the lion than you were," Ming Li said. "You have to admit, it had weird sparkling eyes. I think your song is about this place. I feel certain about it. Now we have to find it again."

"So now, instead of trying to get through this, we're going to search for a lion?"

"I don't think we're getting anywhere, so if we find anything, I think we're okay."

Wren blew out a slow breath. "I wish we were with Graham and Tal. I hope they're okay." She hoped Ben was okay as well, but he could take care of himself. If he wanted to, he could fly out. They turned a corner, and she sighed. This path looked like it went on forever. They had been walking all day, and the moon was rising, casting a soft silvery glow over the maze.

Ming Li glanced sideways at her. "What if you could only find one of them?"

"I'm not going there," Wren said, shooting her what she hoped was a look that would make her drop it. No way was she having this conversation in front of Sen.

"Come on!" Ming Li pled. "Check out this pathway. It's going to take us forever to walk there. It's going to be so boring."

"I want to find both of them."

"I'm guessing if you could only choose one, you would choose Graham now that you know Tal has an arranged marriage."

Wren clenched her fists. "We're fifteen! I don't have to choose anyone! Why do we even have these conversations?"

Ming Li shrugged. "Because they're fun."

Wren stopped and grabbed Ming Li's arm, causing her to turn and face her. "They aren't fun. They're embarrassing and I hate them."

"Why embarrassing? It's a hypothetical question."

"You don't get it, do you?" Wren dropped her friend's arm and pushed a piece of hair behind her ear. "It's one thing when we're alone, but it's way worse when you do it in front of other people."

"No one's here but Sen, and he doesn't care," Ming Li said, motioning to Sen.

"I really, really do care," Sen said, leaning against the hedge and focusing on his feet. "These kinds of conversations are painful."

"They are not."

Wren shook her head. "They aren't painful for you because you're not the one getting asked all the embarrassing things."

"You can ask me anything. You should know that. Questions are just questions. That shouldn't be embarrassing."

Wren put her hands on her head and walked in a slow circle. She didn't know how to make Ming Li understand.

"Are you gonna have a meltdown? You look like my mom does when she's about to have a meltdown."

"New rule," Wren said in a stern voice. "You are not allowed to talk to me about boys if anyone else is within hearing distance. Alright?"

Ming Li studied her face for a moment, then sighed. "Okay fine."

"Did anyone hear that?" Sen asked, searching the sky. They all looked up and heard a loud screech, but Wren couldn't see anything.

"What was that?" Ming Li wondered.

"There!" Wren pointed at a shape flying above them. Whatever it was circled them like a vulture.

"Run," Sen said calmly. "Go."

They all took off in the same direction. Wren glanced over her shoulder and saw something flying after them. A stone dragon with two horns coming out of its head dove towards them. It looked like a statue off of an old building. It opened its mouth and shrieked. Wren screamed and focused on moving forward. Ming Li and Sen were barely ahead of her, but she was keeping up.

"Get off!" Wren yelled as the creature clawed at her hair. She waved her hands over her head and hit something solid. She dropped to the ground, and it flew over her. Getting into a squatting position, she watched it chase after Ming Li and Sen. She took three deep breaths and took off after it.

"Ahhhhhhh!" Ming Li screamed as the creature grabbed her cloak and yanked her to the ground. After she fell, it ignored her and kept following Sen. Wren jumped over Ming Li and kept running. She unfastened her cloak and pulled it off. Right as the beast dove at Sen, she whipped it with her cloak. It pulled up and turned towards her, landing on the path. She clutched the cloak and breathed hard.

"Hey! This way!" Sen called to the creature. It turned and squawked at him.

"No! Over here!" Wren yelled. It glanced back and forth between the two of them. She could sense Ming Li moving behind her, but she didn't dare to take her eyes off the monster. Its head tilted in confusion. Its expression didn't change, but she assumed it probably couldn't if it was actually made of stone.

"Careful," Ming Li said, quietly coming up to Wren. The monster took a step towards them.

"Here!" Sen called. When it turned its head to inspect Sen, Wren grabbed the edges of her cloak and snapped it over the beast's head.

"Run, run, run!" Wren yelled, dodging the struggling monster and following Sen. She could hear her cloak being ripped to pieces. They were almost at the end of the path when they saw the lion run past.

"Follow the lion!" Ming Li yelled, right before they collided with Graham and Tal. The five of them struggled to untangle themselves from one another and jumped to their feet. Ben was watching all of them, with his head tilted to one side.

"We need to keep up with the lion," Graham puffed out of breath.

"Okay, let's go." Wren picked Ben up and sprinted after the animal. They all raced together, not letting the lion out of sight. He made a bunch of turns, but they stayed close enough to see where it was going. When they rounded a corner after him, they came into a large hedge circle. The lion was nowhere in sight.

"Where'd he go?" Tal asked, breathing deeply.

"Are we in the center?" Wren asked, walking in a circle, looking up at the tall leafy walls. The area they were in was at least thirty feet around.

"Where did the entrance go?" Ming Li asked. Wren spun around. They were absolutely blocked in.

"Now what?" Graham asked, observing the circle. "This hedge is taller than everywhere else."

Sen shrugged. "We can still climb it."

"We should wait until we're sure the gargoyle-y thing is gone," Ming Li said. "I don't know how to fight against stone. There aren't any things in here to use as weapons."

Graham wiped the sweat from his brow. "Gargoyle?"

Before anyone could respond, the hedge parted, and the lion stepped out.

"Don't stare into his eyes," Wren cautioned, staring at its massive chest.

"Why?" Tal asked.

"The song says, 'Do not let it hypnotize,' so I'm assuming that would be through his eyes."

"Riiiight," Tal said. "I couldn't remember the whole thing."

"Why are we assuming it's a he?" Ming Li asked, shifting nervously.

Tal yawned. "Because he has a mane."

Graham shoved Tal. "Are you staring into his eyes?"

"Thanks. I slipped into that way too easily," Tal said, shaking his head.

"Look!" Wren said, pointing. The hedge was pulling away to reveal a person lying on the ground. He had blond hair and was turned away from them, with his cloak covering most of him. The hedge pushed him and he rolled to face them.

"It's Jaaz," Graham said, approaching the sleeping boy. The lion growled and stepped in front of him. Graham stopped and glanced helplessly at his friends.

"Who's Jaaz?" Sen asked.

"That's the guy I told you about," Ming Li said. "He betrayed us at Meegore. He was working with The Dark Cloud."

"Look," Tal said, pointing up at the sky.

Wren gazed up, hoping it wasn't the stone dragon. A fairy was flying in large circles, slowly getting lower. With delicate grace, she landed on the lion's head and glanced at them all with an amused gleam in her eye. Her pink dress glittered, and in the moon's glow it almost looked like it was covered in stars. The night breeze blew the fairy's long blond hair softly around her face. Wren wasn't sure what to think. Vork seemed to believe they shouldn't be trusting fairies. The question Wren had was, should they trust Vork?

"You have reached the center of the maze," the fairy grinned. "Congratulations. Now, what do you desire?" They all looked at each other and from Jaaz to the fairy.

Wren stepped forward. "We've come to get a golden tear."

The fairy nodded. "I thought that might be the case. You have a choice to make. You may have a golden tear, or you may free this boy," she said, pointing at Jaaz. Jaaz groaned and sat up. He gazed around, squinting at everyone.

Wren peered at Graham, and his eyes held the same confusion she felt in her own.

"We need that tear," Ming Li said.

"We can't leave him here," Wren said, gesturing to Jaaz.

Ming Li tilted her head. "Why not? He's a traitor."

Graham scratched the back of his head. "Yeah, but he doesn't remember that."

"Actually I do," Jaaz said, standing up.

Tal narrowed his eyes. "But Brake erased your memory."

"I woke up at home and I couldn't remember anything from the last two months," Jaaz said, focusing on his feet. "My dad questioned me a million times. Other people came and asked me questions. I couldn't remember anything. About a week ago, I saw Fleetman Zera, and it all came back to me. I came to find you all and let you know how sorry I was, but all I could remember is someone mentioning that we had been in the jungle. I spent days in the jungle and then one day, I was captured by a stone dragon and dumped here."

"So we can leave him, because he remembers," Ming Li said.

"I am sorry," Jaaz said, pulling his cloak around himself. "I know I was being ridiculous. I've thought a lot about my life and all the dumb choices I've made."

Ming Li crossed her arms. "Of course he's going to say he's sorry."

"If it's more important to get the tear, I understand," he said, kicking at the dirt.

Graham sighed and studied his friends. "We've all done things that are stupid. It would be terrible if people never forgave us and we never had a chance to become better."

Wren nodded. "That's true. If we never forgive people, we'll become bitter."

"What will happen to him if we don't take him?" Tal asked the fairy.

"He will eventually become part of the maze," she said, shrugging with indifference.

Ming Li looked at Sen. "You're with me, right? We have to have that tear for the silver eclipse. Without it, the weather might never return to normal."

"I don't know," Sen said, not meeting her eye. "Can you leave here with a clear conscience, knowing he won't ever get out?"

Ming Li bit her lip and stared at Jaaz. He kept his gaze on the ground. She stared at him for what Wren felt like was an eternity, and then a tear ran down her cheek. "No. I can't," she sobbed. "I'm sorry I'm the way I am." Wren put Ben on the ground and put her arm around her friend and let her cry quietly on her shoulder.

"We choose him," Graham said, pointing to Jaaz. The fairy nodded and flew away. Jaaz sank to his knees and buried his face in his hands. His shoulders shook as he cried.

"I thought you would leave me here," he sniffled. Wren wiped a tear from her eye and walked over to Jaaz. She stood beside him and placed her hand on his shoulder. He jumped in surprise and looked up at her with red eyes. "Thank you."

She took his hand and helped him stand. Ming Li wiped her nose on her cloak and walked over to where they stood. "Don't make us regret it," she sniffed, pulling the tall boy into a hug.

His eyes widened, and more tears ran down his face. "I won't."

Something weird was going on in Graham's throat. He must have swallowed funny, because there was no way he was going to cry.

"Sorry," Jaaz said, wiping at his face, "I don't remember the last time someone hugged me."

Graham felt his bottom lip tremble, and he forced himself to look away as Wren joined the hug. Why was he about to cry? If anything, he should be upset that they didn't get the tear. This was the closest he had ever come to crying and feeling happy at the same time. He needed to get better sleep.

"Anyone can join in," Ming Li said. Graham shook his head as he watched Wren and Ming Li.

"I'm good," Sen said as Tal walked over and clapped Jaaz on the back.

The lion roared, and they all jumped. Graham had completely forgotten about it. It walked towards them and stopped. He closed his eyes and a golden tear fell to the ground. He bowed his head to them and disappeared into the hedge. Graham bent and picked up the solid tear. He turned it around in his hands and held it up as his friends gathered around to see.

"That is so cool," Ming Li said, touching it softly.

"That's the biggest tear I've ever seen," Tal said, looking closer. "It's a good two inches."

"Why don't you keep it," Graham said, handing it to Wren. "I don't want to lose it."

Wren took it and put it in her pack. "It's so neat we're making progress. It's feeling like we might actually find everything for the silver eclipse."

"I hope everything we need doesn't take as much effort as the first two things."

"It feels like I've been in danger a lot since I met you guys," Sen said.

"Do you guys know the way out of here?" Jaaz asked. "I got dumped here, so I won't be helpful."

"This place doesn't make any sense," Graham told him. "The hedges change, so I don't think there's an exact path. Wren should probably grab Ben. He doesn't have the best sense of direction and if we run after him, we'll end up lost."

Tal laughed. "Wren and Ben. I never thought about that before. And now we have Sen, Wren, Ben."

Wren rolled her eyes. "The E-N names were big for a long time." She picked Ben up and walked to the opening. "Are we going?"

Ming Li nodded. "I'm ready. Graham, you should pick some berries and use them in your nasty medicines."

"That's not a bad idea," he said, plucking some berries and dropping them in his pack.

"I already picked a bunch for you," Wren told him, patting her bag.

"Look," Sen said, pointing ahead. "It looks like the hedge goes on forever and then opens. I think it's letting us out."

Ming Li playfully shoved her shoulder into Sen. "Last one out has to clean up the meetinghouse for a week!"

Sen's eyes sparkled. "Ready? Set? Go!" Everyone took off running.

"This isn't fair!" Wren called from behind. "Nobody else is carrying a dragon!"

Graham stopped and turned around, holding out his arms. "I'll take him." He smiled at the look of disbelief on her face. Her eyebrows almost popped off her head when Ben jumped from her arms and into Graham's. "We bonded."

"That's so sweet," Wren said, rubbing Ben's head. Her eyes twinkled mischievously, and she turned and ran.

Graham shook his head and glanced at Ben. "Should we let her think she's going to beat us?" Ben stared at him. "Alright, let's go." Graham tucked Ben under his arm and ran. Sen, Ming Li, and Tal all had a big head start, and they were together, with Jaaz slightly behind.

Wren wasn't far in front of him, but he figured he would let her stay ahead for a minute. She had gotten pretty fast. The Wren of today could run circles around the Wren of five months ago. She turned her head and smiled. Her eyes flashed with excitement. Graham felt his heart melting. That didn't mean he wanted to lose, though.

He sped up, so he was even with Wren, and grinned mischievously. "Hey. How's it going?"

"Ugh," Wren said, focusing her eyes forward and going faster. She might have gotten faster, but Graham had been running his whole life. He sprinted ahead, and as he passed her, he turned and winked. He wasn't paying attention and slammed into the hedge, falling onto his back.

"Ouch," he muttered as Ben flew a few feet off the ground and growled at him.

"Are you okay?" Wren asked as Ben landed in her arms.

"Fine," Graham said, sitting up.

She waved. "Good. See you at the end!"

"That was faster than we expected," Brog said, as they all gathered outside the maze. He sat near a small campfire with Austra and Padmire. "Did you get it?"

"Yes," Wren said, holding up her pack. "But we only got one."

"I assumed that would happen," Austra said, pushing her messy black hair over her shoulder. Wren's mouth turned down at the sight of her. The spotless dress she'd been wearing the day before was dirty and ripped, and she had a streak of dirt running across her cheek.

"What happened to you?" Tal asked.

"I told Vork you would only get one tear and we might as well fight now to save time."

Tal looked around. "Where is he?"

Austra sneered. "He lost and ran off to lick his wounds."

"It was the craziest fight I have ever seen," Brog said, shaking his head. "I offered to fight him, but Austra was determined to do it on her own."

"How do you fight with a goblin?" Wren asked.

Austra shrugged. "Well, we started out fencing, and then it turned into a bit of a wrestling match."

"I am having the weirdest image right now," Sen said. The corners of his mouth twitched, but he didn't smile.

"It was the best fight I've seen in all my years," Padmire said, from his place on Brog's shoulder. "I wouldn't suggest making this woman angry," he said, pointing at Austra.

Austra crossed her arms and raised an eyebrow. "Sound advice."

"Why is Jaaz with you?" Brog asked.

"I think he was brought to the maze to test us," Wren said.

Ming Li studied the ground. "Some of us almost didn't pass."

Jaaz kicked at the dirt. "I wouldn't have blamed any of you if you left me."

Austra drummed her fingers against her arms. "So, what do we do with him?"

"Do you have to do something with me?" he asked, gazing hopefully at Austra.

"We might need to have Brake erase his memory again."

"Please don't!" Jaaz begged. "That messes with a person."

"He remembers everything," Graham said. "It all came back to him. I don't think he's going to be a problem again."

Austra kept focusing on Jaaz for a moment and sighed. "Fine. For now. But he doesn't get to know anything."

"Phew," Jaaz said, shaking his blond head. "There are some memories I've had the last few weeks that I don't want to forget."

Ming Li raised her brow. "I would think getting captured and dumped in a maze wouldn't be the best memory to keep."

"That's not the memory I mean, although I appreciate the memory of you all sacrificing for me. I know I don't deserve it."

"So what glorious memory have you had?" Austra asked, narrowing her eyes.

"My dad secured me an arranged marriage."

Ming Li snorted. "Anyone we know?"

"Oh yes," Jaaz said, with the biggest smile Wren had ever seen on him. "Why don't you guess?"

"Now isn't the time," Austra said, clenching her teeth. "There is nothing redeemable about arranged marriages. If I could run for the Fleet, that is the first thing I would lobby against."

"Why can't you run?" Graham asked.

"Because I was born on the other continent. It's against the rules."

Tal's lopsided smile appeared. "I really want to know who this girl is."

"Me too," Graham added.

Austra tapped her foot. "Fine. Make it fast."

"I had a fight with my dad," Jaaz explained. "He tried to get me to help The Dark Cloud again. I told him I wouldn't. We argued for a while and I told him he wasn't as important and influential as he thought he was. He told me he could prove I was wrong. He said I could choose any girl I knew and he could get me a marriage contract. I chose the hardest one I could think of. The next day, he came and told me it was arranged. I never in a million years thought it could happen."

Ming Li threw up her arms. "Just tell us who."

"Don't get mad at me, okay Tal?" Jaaz said, with shining eyes.

Tal tilted his head quizzically. "I don't have any sisters, so I don't see why I would care."

"It's Solia."

Wren covered her mouth to suppress a laugh, and Ming Li giggled. Graham seemed amused and Sen, Brog, and Austra looked indifferent.

If Tal could smile any bigger, his face would probably crack. "Why would that make me mad? That's the greatest thing I've heard in a while. And how does Solia feel about this?"

"She doesn't actually know. Her parents thought it would be a good idea to not tell her, since she hates me and all."

Tal's eyes were shining. "So, when is she going to get to know? On your wedding day?"

"Hopefully before that, since that's not for seven years. Her dad wants me to try to get her to tolerate me first."

Tal burst into laughter. "Good luck. Solia doesn't like anyone."

Jaaz shifted his weight from one leg to the other. "That's not completely true. She likes you."

"Not really," Tal said. "She likes that my dad is the governor. How did your dad get her dad to agree?"

"I'm sure it has something to do with money. My dad has a lot. Maybe you can help me get her to like me?" he said, gazing hopefully up at Tal.

"I'm fairly certain this conversation can wait until a different time," Austra said, fastening her cloak. "Preferably a time when I am not present. Where is Vork? We need him to send us back. Vork!" she yelled. "I know you're here somewhere! Get us out of here!" The ground rumbled and everything went dark. When Wren could see again, they were back at the bottom of the snowy mountain.

Wren hugged herself against the blowing snow. It was so cold! She hadn't missed her cloak until now.

"Grab my cloak before Wren freezes," Sen commanded. They all grabbed hold and swirled away.

# Chapter 14

"This seems like a huge waste of time," Graham said, lounging in Blake's sitting room. "It's been a week since the maze and there are so many other things we should work on."

Tal grinned. "Come on. We're doing this for a friend."

"What if Brake comes home and wonders what we're doing here?"

"Technically, this is your house too now, right? I mean, Brake is your dad."

Graham sighed. "I still don't think he would consider this the best use of our time."

Tal threw out his arms. "Come on. What could be more fun than getting Solia to fall in love with Jaaz? Don't you wanna be part of that story? It may go down in history as some bizarre fairytale. The geeky boy falling in love with the beast. That actually sounds familiar."

"How do I look?" Jaaz asked, entering the room. He held up his arms and turned in a circle.

"Your green tunic brings out your eyes," Graham said, trying not to smile.

"Really?" Jaaz asked, inspecting his shirt.

"I don't know, man. This isn't my thing."

"You look fine," Tal assured him. "You look a lot better with a longer tunic. Stop wearing the short ones. They make you appear awkward. You're a growing boy. You can't be wearing last year's tunic now, can you?"

"It was actually from two years ago," Jaaz admitted. "I don't care what clothes I'm wearing so long as I'm not naked."

"Well, from now on you care," Tal said. "No more outgrown clothing, and never wear your shirts wrong side out."

He shrugged. "I can't remember to think about stuff like that. I haven't worn my shoes on the wrong feet for over a year, though. That has to be a plus, right?"

"Riiiight. Here," he said, walking over and mussing up his hair. "You don't want helmet hair. Let it breathe a little. Get some gel or something and run a bit through. And while we're on the subject, if your clothes get dirty, you change them. A girl doesn't want to see what you've been eating for breakfast."

"How do you enjoy life if you have to think about so many things?"

Tal grimaced. "If you keep at it, it will become second nature and you don't have to try so hard."

Graham leaned back in his chair and grinned. "He's only wearing one sock."

Tal put his hands on his hips in true Ming Li fashion. "Where's your other sock?"

"I'll be wearing boots so nobody will know. When I was changing, I needed a handkerchief and I couldn't find one so–"

Tal put his hand up. "Just stop right there. We don't use our socks as handkerchiefs."

Graham turned a laugh into a cough and stood. "I'll go find him some." He sprinted upstairs into his room and glanced around. It had been a long time since he'd been here. He hadn't lived here long, so it didn't actually feel like his room. He opened a drawer and pulled out a pair of socks and hurried back. When he entered the room, Tal was rubbing his temples with his fingers.

"No, you can never use a string from your cloak as floss," he was saying. Graham hid a smile and threw the socks to Jaaz.

Jaaz caught them and sat on the floor. "It's no wonder you all seem stressed all the time." He pulled on a sock. "It must be hard to try to follow so many rules."

"Okay now, you'll probably only have a few minutes, if not seconds, to talk to Solia," Tal instructed. "She's going to be leaving her music lesson and walking home. You want to make the time count."

"So I hurry and ask her out? That's never gone well for me."

"No, don't ask her out this time. We want to start small so we don't ruin anything."

"Just talk to her," Graham added. "Ask her how she's doing. It doesn't have to be anything big. Tell her she looks nice or something, but don't be creepy."

"Is this what you two do? Are either of you finally dating Wren? I could tell you both wanted to when we were at Meegore."

Graham folded his arms and cleared his throat. "We are kinda busy these days. We don't have a lot of time for dating."

Jaaz scratched his head. "Hmm, should I be taking your advice when you don't even have girlfriends?"

"There's no point in dating for me," Tal muttered. "I have an arranged marriage and no, I'm not talking about it."

"What about you, Graham? Do you have an arranged marriage?"

"No."

"So why don't you date?"

"Who says I don't date?"

Tal grinned mischievously. "It only counts as a date if the girl knows it's happening." Graham took a pillow off the sofa and flung it at Tal's head. He caught it and laughed.

"I kind of figured you would both date all the time. All the girls at school like both of you," he said, rolling his eyes. "Most girls don't even know I exist."

"I'm sure that's not true," Graham said, sitting back down.

"Wren didn't even remember me and she fell on my head."

Tal shrugged. "Well, that was the old you. Stand up. See? Just wearing clothes that fit you and changing your hair a little makes a tremendous difference."

"You think so?" Jaaz asked, straightening his shirt.

"You need to remember that you can't get discouraged if Solia is mean to you. It's kind of part of her personality, so it's probably going to happen."

"What if I don't know what to say?" he asked, fiddling with his hands.

"We'll be close by. If you sound like you need help, we will send you a message," Tal assured him. "A brief message so we don't give ourselves headaches."

Graham nodded. They had learned the hard way the consequences of sending messages to each other's minds. If you talked too long, it came with a headache and the occasional throw up.

Jaaz tied on a black cape and nodded. "Okay. I think I'm ready."

"So this is Boztoll these days," Wren said, walking down a foggy street. It wasn't terribly dark, but the sun was behind the clouds.

"It's foggy most of the time," Sen explained. "About once a day Valeena parts the clouds and everyone runs outside, but it only lasts about thirty minutes."

"Is it hard to do?" Ming Li asked. "Couldn't she do it more than once?"

Sen shrugged. "It takes a lot out of her."

"Why don't they repair their houses?" Wren asked. "At least it would make the city look better. A coat of paint would do wonders."

Ming Li nodded. "And some new roofs. Look at the way the shingles are curling up on most of them."

Sen glanced sideways at them. "It doesn't go well for anyone in Boztoll if they appear too prosperous. People here have learned to mind their business and, as my dad says, fly under the radar. We know there are members of The Dark Cloud living here and they don't cause problems unless things get too cheerful."

Wren clenched her fists. "It makes me so mad! Not only do they make people move here, but they don't even let them live normally."

"The Dark Cloud is all about power and control. Their motives make little sense," Sen agreed.

"I'm surprised they let Valeena part the clouds at all," Ming Li said.

"We don't know who they are," Sen explained. "All we know is they're here. We think they let her do it so everyone doesn't leave. If they stopped her, they would all blow their cover. Here we are." The house they approached was about as nondescript as possible. A simple gray square with a white door. The paint wasn't new, but it wasn't peeling like most of the other homes. Sen knocked.

The chill in the air had Wren pulling her cloak closer as they waited. While reading about the next step to the silver eclipse, something reminded Sen of a remark Valeena had made, which was why they were here now.

The door opened a crack and someone cautiously peeked out. It closed, and they heard the chain being taken off. They filed into the house and the door shut behind them. When Wren turned to see Valeena, she saw her dad instead.

"Dad! What are you doing here?" she asked, throwing herself into his arms.

Sen scanned the room, frowning. "Where's Valeena?"

"Valeena was needed in Akkron," Drew said, keeping his arm around Wren. "I tried to stay hidden on the other continent, but it's too different over there. I don't know the customs and how things work. It felt like a lot of effort, so I came back. I seriously doubt the governor is searching very hard for me, and he would never think to look here."

"Does Valeena know you're here?" Sen asked.

"Of course. She has some events she needs to attend with her husband, so she told me I could stay here until she comes back. It sounds like she won't be back for at least a month. It's a busy political year. Come sit," he said, leading them to the kitchen. They sat at the small square table and watched as Drew cut them all a piece of cake.

"This is Sen," Wren told her dad.

"Ah, yes," he smiled, placing a large piece of cake in front of Sen. "Valeena told me all about you. I'm glad we finally have all the pieces coming together. Now that you've all been in the cave, big things should be happening." He gave them all some cake and sat with them. "Why are you looking for Valeena?"

"We've been collecting things for the silver eclipse," Ming Li told him. "Sen thought Valeena might be able to help with one of the instructions. There's no way my mom made this," she said, chewing quickly.

Drew laughed. "Boztoll doesn't have a bakery. I made this, so it's not going to be like Mali's. Sorry it's disappointing."

Ming Li shrugged. "It's fine. Way better than anything Brake could make."

"Well, the day I cook something worse than Brake is the day I stop cooking." They all laughed.

"I picked a ton of really good berries," Wren said, putting her fork on the table. "I'm hoping Graham can use them in his medicine because his medicine is as bad as Brake's food."

"That's a good idea," Drew said. "Brake should try cooking with them. It couldn't hurt, right?"

Ming Li nodded. "Now that Zera and Brake are inseparable, Brake hasn't been cooking anything. Zera's food is tons better."

"Zera and Brake have been inseparable?" Drew asked with wide eyes.

"Oh right. You missed all of that," Wren said, wondering how she had forgotten her dad hadn't been around. "I don't even know where to start."

Ming Li leaned back in her chair. "It's not that complicated. We know about all you guys' younger days fighting bad guys and stuff."

Drew scratched his chin. "Brake told you?"

Wren nodded. "It all came out when we found out Graham is Brake and Zera's son."

"What?" Drew asked, all the color draining from his face. "Graham is their son?"

"Yeah," Ming Li said. "It was all kind of crazy. Brake found Graham's shirt that said, 'Dryson' on it and they figured it out."

"Graham Dryson," Drew whispered, rubbing his forehead. "I should have known. I thought Brake's baby was named Bryson." He stood up and walked to the wall. He put his fist on his head and leaned it against the wall. "If I wasn't so bad with names, I could have told Brake immediately."

Wren walked over to her dad and put her hand on his back. "It's not your fault, and it all worked out. Graham hasn't been here that long. It just feels like it because everything is so crazy lately."

"I suppose you're right," he said, turning to Wren. "I should be happy. Brake and Zera were so heartbroken. I watched them suffer for years. Now they can try to put it all behind them and start over. Graham's a good kid. I think it will work out."

"It will," Wren agreed.

"Where is Graham anyway? And Tal?"

"Now that's a weird story," Ming Li said, standing and leaning on her chair. "They're with Jaaz, trying to figure out how to make Solia fall in love with him."

Drew blinked twice. "Solia? The girl who is always mean to you?" he asked Wren. She nodded. "And Jaaz? Why are they helping him?"

"It's a long story," Wren said. "And I'm pretty sure it's going to have a bizarre ending."

"Here she comes!" Tal whispered, ducking behind the large moakberry bush Graham was squatting behind. "Remember, we can focus on you and whisper, and whatever we say will come into your head. You can do the same back, but try not to, or she might think you're crazy."

"I remember," Jaaz said, giving him the thumbs up sign.

"Okay," Graham told Jaaz, "Now sneak over to that tree and walk down the path like you're coming around from the other direction and you have to pass her."

"Right. Stealth mode," Jaaz said, bending at his middle and walking quickly behind the bushes and around a large tree. When he came to the tree, he stood up and walked onto the path, giving the illusion of coming around the corner. It would take a minute before they passed each other.

"What is he doing?" Graham whispered. "He's walking like a dork. Why's he swinging his arms like that? And he's not bending his knees."

"I guess we should have practiced," Tal muttered under his breath. He closed his eyes and whispered, *Jaaz, walk normal.*" Jaaz stopped for a second and started walking again, this time a little less oddly.

"She's seen him," Graham said. "She rolled her eyes. Not a good sign."

"This isn't going to be a simple thing. Solia's a brat. It's going to take time."

"I hope she isn't too mean. I don't know if he can handle it."

They were going to pass any second. Solia's mouth was puckered like she had sucked on a pickle. She held her music close to her and moved as far to the left as she could without going off the path. She appeared irritated that Jaaz would dare walk on the same planet as she did. Jaaz had caused a lot of problems, but Graham was sure he deserved someone better.

"Is he going to talk?" Tal mumbled. It seemed like they might pass with no interaction. At the last second, Jaaz stepped in front of Solia and they collided. She rocked back and dropped her music on the ground, grabbing Jaaz to steady herself.

"What do you think you're doing?" she demanded, shoving him back a few steps.

"Hi, how are you today?" Jaaz bowed.

"Oh boy," Graham said, smacking his head.

"What do you mean, how am I?" she asked, tapping her foot on the dirt. "You nearly knocked me over. Watch where you're going!"

"I'm doing pretty well myself," Jaaz said.

"Are you crazy? Pick up my music and don't smudge it."

"Okay, thank you," Jaaz said, squatting to retrieve the fallen papers. "By the way, your hair is the color of baby ducks. Fluffy ones that, you know, are still cute. Oh, and are yellow. Some baby ducks are brown and some have spots. I think I saw a tan one once. But your hair is like the yellow ones."

Tal covered his mouth and snorted. His shoulders shook as he tried to control his laughter.

"Shhh," Graham shushed him. "She's going to hear you." Tal kept shaking. *Jaaz, give her the papers and leave,* he whispered.

Solia was staring at Jaaz with her brow furrowed and her mouth hanging half way open. He held the papers out and she slowly took them.

"Maybe I'll run into you some other time," Jaaz said, still blocking her path.

"Please don't," she said, walking around him.

*"Let her go,"* Graham whispered.

*"No, I'm gonna try again,"* Jaaz whispered loudly.

"What did you say?" Solia asked, whipping back around to face Jaaz.

"I didn't say anything," he said.

"Yes, you did. You whispered something," she said with narrowed eyes.

"No I didn't."

"I heard you."

"Oh. Well, I wasn't communicating with people in the bushes. I was talking to myself."

"Oh, come on, man," Graham said, shaking his head.

*"Jaaz, abort the mission. You are sinking,"* Tal whispered, still smiling.

"You know, your shirt looks really blue," Jaaz said.

"Do I know you?" Solia asked, glaring at him. "Why are you even talking to me?"

"We have magical equality together. Last year we had math and gym. The year before, I think it was history. Five years ago you punched me in the nose and took my pencil. It bled, but not too much. My mom wanted to talk to your mom, but my dad said not to get involved."

"*Get out!*" Graham whispered.

"*Now,*" Tal agreed.

"*Run, you won't be able to recover if you keep talking.*"

"*He's not leaving,*" Tal said in Graham's mind.

"*I know. What do we do?*"

"*We should leave him on his own. He's not listening.*"

"*I was thinking of tackling him.*"

"*Why are they just staring at each other? Do you think she's going to punch him?*"

"Whoa," Graham said. "Why are we talking to each other in our minds?"

"Oh man," Tal whispered. "I might puke."

"Never talk to me again," Solia finally said.

"But I'm planning on asking you out in the near future. Does that change anything?"

"Don't ever ask me out," Solia said, poking him in the chest with her pointer finger. "I am way too popular for you."

"My dad has more money than your dad."

"Excuse me?"

"Your family is all about money, and my family has more money."

"*Stop talking!*" Graham whispered to Jaaz. This kid didn't know when to quit.

"Your family has more money than mine? Please," she said, turning and walking away.

"You can ask your dad!" he called after her.

Graham sat cross-legged in the dirt and rested his elbows on his legs. He put his face in his hands and counted the pulsing in his head. He heard Tal puke in the bushes. Someone needed to perfect this way of communicating.

"Hey, guys. I'm done," Jaaz said, bounding up beside them. "It went pretty well for a first try."

Graham gazed up at him through squinted eyes.

Tal wiped his mouth on the back of his hand. "You think it went well? It was awful!"

"Which part?" he asked, sitting on the ground.

"The beginning, middle, and end," Tal said.

Graham took a medicine bottle from his pocket and took a swallow, and handed it to Tal. "Why did you bump into her?"

"I wanted it to look like an accident so she wouldn't think I was talking to her on purpose."

"It didn't look like an accident," Graham said, as his head began to clear. "You stepped right in front of her."

"Yeah, but it worked."

Tal shook his head. "If she ever sees you again, she's going to run."

Jaaz sighed. "Why are girls so hard?"

"It's alright," Graham said, clapping him on the back. "We'll make sure we practice more next time. Maybe help you with some better lines."

"I don't know," Tal smiled. "I'm not sure if we can beat the duck line."

"You liked that?" Jaaz said, bouncing. "I thought it was pretty good."

"Next time you need to listen to us," Graham told him, getting up off the ground. "My medicine works fast, but that is not a pleasant thing to go through. If we tell you to stop, you stop. Alright?"

"But think of what I would have missed if I stopped today when you told me to," he said, tilting his head.

Graham crossed his arms and frowned. "Believe me, I am thinking about it. If you don't agree to listen, I'm not helping next time."

Jaaz nodded. "Alright, fine."

"I had a great idea," Tal said, standing up and brushing off his pants. "Oh yes. That's a great idea! I'm pretty sure I'm a genius. There is no way she will be able to resist talking to you."

Jaaz stood up and leaned towards Tal. "What is it?"

"Oh, it's too good to tell you. I'm gonna have to show you."

# Chapter 15

"Wren, wake up!" Ming Li said, shaking her shoulder.

Wren rolled over and sat up. Ming Li was kneeling at her side. "What's wrong?" She rubbed her shoulder. It was sore from sleeping on the floor of Valeena's house.

"There's a huge storm. I can't believe you can sleep through this."

Now that she was awake, Wren could hear the rain beating on the house, followed by the deep rumble of thunder. "Wait, you woke me up because the storm didn't?"

"I peeked out the window and there were cloaked people walking around! My guess is they're Dark Cloud. I think they're behind the storm."

"What can we do?" Wren asked, fully awake now. She pulled on her boots.

"I don't know," Ming Li said, putting the hood of her blue cloak over her head. "We have to do something, though."

"Where's Sen?" Wren asked, glancing over to the corner where a mess of blankets lay empty on the floor.

"He must have gone out." Ming Li muttered, handing Wren her cloak. Wren accepted it and stood.

"I assume my dad is still sleeping. He's the hardest person to wake up."

"I guess that's where you get it. It wasn't easy to wake you. Let's go."

Wren covered her head with her cloak and made sure it was fastened. She followed Ming Li out into the wind and rain.

"This is bad!" Wren yelled into the wind as her cloak whipped against her and her hood blew off her head. Long tendrils of hair slapped her in the face. She grabbed a handful of hair and tried to tuck it into her collar. Lightning flashed and a loud crashing sound shook the ground.

"I think it hit something over there!" Ming Li called to her. She turned and sped down the paved street.

Wren bowed her head against the wind and followed. They ran against the wind, hair and cloaks billowing behind them. They dashed around a corner and Wren could see fire ahead. As they came closer, she realized it was a store. A small crowd had gathered and was watching the building burn.

"The lightning must have hit it!" Ming Li yelled.

"We need Graham. He might be able to put it out."

"It's too far gone."

"This has to end!" a man in the crowd yelled. "They send us to Boztoll and then they harass us here!"

Wren scanned the crowd of people. Some of them appeared angry, but most of them looked worn and tired. Lightning was flashing all around. Wren could feel the electricity in the air. There had to be something they could do.

"Everyone needs to go inside!" another man called. "It isn't safe out here. The rain will put the fire out."

"Come on Governor Calsen," the first man said. "We're starting to think you are as bad as Governor Briggs! You knew the problems Boztoll had when you were elected. You promised to change things!"

"My Fleet and I are working on it. This isn't a problem that can be solved overnight."

A woman sprinted down the street and stopped in front of the governor. "There's a barn on fire outside of town!"

The first man clenched his fist and glared at the governor. "This lightning isn't going to stop! It's too much. Everything will catch fire!"

Wren looked into the sky and everything inside of her called out to the lightning. "Everyone, stand back!" she yelled. She spread out her arms and concentrated on the lightning. Every single bolt felt connected to her. She could sense it before it left the clouds.

"What's she doing?" she heard someone ask behind her. She could hear people talking, but she was focused on the clouds.

She pinpointed every cloud that was about to shower down lightning, and she clenched her fists and pulled at it. At least thirty bolts of lightning came bursting from the clouds, and Wren opened her fists and pointed them at the sky. The bolts stopped mid descent and turned and hurtled upwards and disappeared into nothing.

Wren dropped to her hands and knees on the wet road and breathed hard. That lightning wasn't natural. Natural lightning was so fast she wouldn't have had time to send it back up. She was almost certain she would have been okay if the lightning hit her. She kind of thought it might absorb into her, but that wasn't something she felt ready to try. Wren gazed up through her wet hair and saw all the people staring at her with their mouths open.

"Are you okay?" Ming Li asked. Wren nodded. "Good. Wish me luck."

"Luck?" Wren asked, getting up on her shaky legs. Ming Li raised up off the ground. "Careful! You haven't practiced enough!" Wren warned her friend. Ming Li had levitated herself up into the meetinghouse tree the other day, but this was a whole different matter. Everyone in the crowd watched as she rose higher and higher. It looked like she was trying to go straight, but the wind was pushing her around.

"What is she doing?" Sen asked, coming up behind Wren.

Wren was wringing her hands. "I don't know." Ming Li disappeared into the clouds and Wren could see the crowd holding their breath.

A woman pointed into the sky. "Look! Something's happening up there." The clouds were blowing away. A minute later, they could see Ming Li, high in the air, with her arms spread out at her sides. When the clouds were gone on one side, she turned to the other side. She did the same thing for the two remaining sides. Wren had never seen her make such a wide wind before.

Everything was brighter now, and the moon and stars were shining through. Ming Li slowly descended until she was standing unsteadily next to Wren. Everyone surrounded them and clapped.

"Who are you?" Governor Calsen asked with wide eyes.

Ming Li swayed and would have collapsed if Sen hadn't caught hold of her. He put his hands on her shoulders to keep her steady. She seemed pale, but it could have been the moonlight. Her hair was one big tangle. Wren wondered if she looked the same. It was taking everything in her to stay up.

"Move! Out of the way!" A familiar voice called. The circle broke open, and Coach Zalliah pushed her way through until she was standing in front of them. "Are you alright?" She pushed back her black hood and her long brown hair blew in what was now only a gentle breeze. This was a weird place to see her gym teacher.

"I need to go lie down." Ming Li said, her head drooping to the side.

Sen scooped her up and turned to Wren. "I'm going to take her somewhere to rest. Don't go back to where we came from." He turned and walked swiftly through the crowd.

"Who are you people?" The governor demanded again.

"Shouldn't you be thanking them?" Coach Zalliah said, glaring at the short, balding man. "Half the city would be on fire right now if it wasn't for them."

"Part of it still is," the governor said, pointing at the almost burned down store.

"But a lot less than it would have been if they hadn't intervened."

"I'm not saying we aren't grateful," the man said. "We are."

"Come on, Wren," Coach Zalliah said, linking their arms and leading her through the crowd.

"Did she say Wren?" someone asked. "Isn't Wren one of the kids that opened the cave at Meegore?"

"You mean part of The Silver Eclipse?" another person asked.

Wren let her teacher lead her away from the group, and they walked silently down the street. She was too tired to ask where they were going. Sen had warned her not to go back to Valeena's house, but she didn't know where he had taken Ming Li. She should have followed.

"Coach Zalliah?" she finally asked.

"Call me Zalliah. I don't suppose you will be returning to school soon."

"Probably not. What are you doing here?"

"Searching for you and the others. You aren't the easiest people to find. You've been busy."

"It's been a crazy year," Wren admitted. "A lot of your training has been helpful."

"I'm sure it has. When I saw the aura around you and your friends, I knew I had to do something. I'm glad I helped you succeed until this point. And the boy with Ming Li, he's the fifth person. It is surprising he came from Boztoll. It's too bad I didn't get to train him."

"He's already pretty good at taking care of himself," Wren said, as the street ended and buildings ended.

Wren stopped and peered into the dark trees. "Where are we going?"

"I need to get you away from Boztoll. All of those people pose a danger to you now."

"Danger how?"

"These people are all living here, unhappily, because a lot of them don't have magic. What's stopping them from capturing you and forcing you to take them to Meegore?"

Wren shrugged. "Maybe the fact that they aren't magic. We want to help them, not become enemies."

Zalliah rolled her eyes. "Yes, that sounds like a glorious dream, but that's all it is. A dream. Now we need to get you somewhere safe."

"What about Ming Li and Sen?" Wren asked, glancing back at Boztoll.

"Ming Li didn't look like she was in any condition to travel. I'm sure that boy will get her somewhere safe. Come, let's go."

"I'm not going to leave them here." Wren said, turning back to the city. Zalliah grabbed her arm, and before she could react they swirled away. "NOOOOOOO!" Wren yelled as they landed at Meegore. "I'm going back," she said, ready to make a portal.

"Wait! Wren." Zalliah said, putting a hand on Wren's arm. "We have colossal problems here. That's why I came for you."

"What problems?" Wren asked, glancing around the quiet island.

"It's the giants," Zalliah said, twisting her hair around her finger. "Come and I'll show you."

"Why don't you tell me? I need to get back and check on Ming Li."

"The Dark Cloud has captured the giants. Every single one."

"How? That's impossible," Wren said, running towards what she hoped was the center of the island. It was hard to know in the dark, with all the trees. This place wasn't exactly easy to navigate in the daytime. She was running slow. Commanding the lightning had been exhausting.

Could the giants really be captured? She saw the crystal ahead and rushed towards it. Off to the side, she saw Haltina, Tnarg, and Mot all tied up on the ground. They seemed to be sleeping. She kneeled next to Haltina and fiddled with her ropes.

"Careful," Zalliah said. "They were given ellebasi serum. It's a lot worse than simple ailam powder. You don't want to wake someone who has ellebasi. They need to wake up on their own or something terrible could happen."

"Who did this?" Wren asked, wishing she had a knife.

Zalliah stared up at the sky and sighed. "I don't think anyone will believe me if I tell them. I saw a man sneaking around in a dark cloak. His hood blew back, and I saw his face ... It was someone I thought was probably dead. I'm even doubting my eyes."

"Who was it?" Wren asked, still trying to untie the ropes.

"Someone you wouldn't know. His name is Gorbin."

"Gorbin?" Wren said, looking up. "We actually have dealt with him."

"So it was him," she said, shaking her head. "There were others, but I didn't see who they were."

"What do we do?" Wren asked, standing. She wasn't going to be able to untie the ropes without help.

Zalliah looked around. "They might still be on the island."

Wren whipped around and checked behind her. "Why didn't you tell me before we came? We should have brought more people. How long will it take for the giants to wake up?"

Zalliah's mouth turned down. "It's hard to say. Some people never wake up from it."

Wren put a hand to her heart. "Never?"

"You can help them," Zalliah said, gesturing at the crystal. "There is supposed to be something inside the cave that will wake someone from almost anything."

"I don't know what that could be," Wren told her. A sense of dread she couldn't understand had started in her stomach and her arms were covered in goosebumps. "There isn't a lot in the cave."

"It was supposed to be a kind of powder or something," Zalliah said, tapping her lip. "The legends aren't clear. If I remember correctly, it was a golden color."

"Golden dust? Wait … there is a bunch of yellow dust there."

"Really?" Zalliah said, with a brightened countenance. "Perhaps there is hope, after all."

"I don't want to go back in there," Wren admitted. "Last time I was stuck for hours."

"Stuck?"

"Yeah, but I guess now I know why. It shouldn't happen again," she said, thinking of Padmire. After the maze, she had sent him back to Meegore. If he trapped her again, at least she could yell and tell him to let her go.

"So you will go?"

"I guess," Wren said, a light appearing in her hand. They walked towards the cave and one side of the crystal glowed purple. The cave opened. "It was good you taught me to climb and stuff. It helps in there," she said, motioning into the pitch back cave. "I still hate going in by myself, though."

Zalliah tilted her head. "I can go in with you if that would help."

"No, that's okay," she said, taking a step into the cave. "We aren't supposed to take anyone without permission."

"Permission from whom?"

Wren shrugged. It might be nice to have Zalliah with her. She was sure to know some helpful magic to get through some places that were tricky. Wren especially dreaded the water. That water was so cold.

"Okay, I'll be here when you come out," Zalliah said. Wren nodded and held up her light. She couldn't believe she was going in here alone again. She took another step, and the cave rumbled shut behind her. "Well, that was a bit unsettling, wasn't it?"

Wren spun around to see Zalliah a step above her. "Why did you come in?"

Zalliah pulled up her own orb of light. "I guess I came too far in. Can you let me out?"

"No. We can't get out until we go through."

"Oh dear," Zalliah said, walking steadily down the stairs. "I guess we'll have to press on now, won't we?"

Wren walked quietly behind her. The hairs on her arms were all sticking up. Something wasn't right. Zalliah had sent them to Meegore the first time by leaving notes in their school books. She had trained them to get over the obstacles inside. When they came out of the cave, members of The Dark Cloud had attacked Graham and Ming Li. Jaaz had betrayed them, or so they thought. What if he hadn't been the only one?

"There's one problem we're going to run into that you didn't help us with," Wren said, not sure where she was going to go with this.

"Oh? What's that?" Zalliah asked over her shoulder.

"There is a place where you have to speak Flordillian. It hasn't been a problem because Ming Li can speak it. I should have learned some."

"No worries," Zalliah said. "Just let me know when it's needed. I speak it fluently."

Wren released a slow breath. "Oh, good." Her heart felt like it was going to bust out of her chest. Could Zalliah be part of The Dark Cloud? One of their members had spoken Flordillian. It could have been Zalliah. She was the one who sent them here. Had she trained them up so they would get her into the cave? Wren kept trudging down the steps without speaking. She had too many thoughts racing through her head. She should probably send a message to the others, but she didn't want to risk Zalliah hearing.

They reached the bottom, and Zalliah scanned the small cavern. Wren wasn't sure what to do. She knew where the latch was to open the door, but should she be helping Zalliah get through?

"There has to be an opening," Zalliah said, shining her light onto the wall.

"It's here," Wren said, opening the door to reveal the two rock slides. Her mind was still racing. Should she send Zalliah down a separate slide or stay with her? If they slid down different slides, she could call her friends and ask for help. They would probably wonder how she had ended up in the cave without them again. Then there was also the chance that Zalliah wasn't bad, and she had been trying to help them.

"Wow, that's dark," Zalliah said, holding her light into the hole. "A bit of a leap of faith, isn't it? So do we go at the same time, or do we need to go on the same side?"

Wren bit her lip. It was time to make a decision.

# CHAPTER 16

"I don't think this is a good idea," Graham said in a low voice.

"You're right," Tal agreed. "It's a *great* idea."

Graham glanced over to where Jaaz was standing with Tal's zebra. "I think we need more time. Solia only had two days to forget about her last conversation with Jaaz. Actually, it's more like a day and a half."

"Do you want to spend a lot of time on this?" Tal asked. "The girls are off working on the silver eclipse and we are playing matchmaker. The sooner we get this taken care of, the sooner we can help."

"You're actually enjoying this," Graham said, looking over at Solia's house. Despite being much smaller than some houses in Akkron, it was still nice. The house had bay windows and a gabled roof, and was two stories tall. It had a meticulous landscape with colorful rose bushes and neatly trimmed grass. The house exuded an overall sense of orderliness and refinement.

"I am enjoying it," Tal said, "But that doesn't mean we don't have more important things to do."

"Okay guys," Jaaz said, leading the zebra over to them, "I think she's starting to like me." He rubbed the zebra's head and it nuzzled into his shoulder.

"Remember," Graham told him, "Don't tell her things that are obvious. Last time you told her that her shirt was blue. She already knew that. You can say her shirt looks nice or her eyes are pretty, but don't state facts."

Jaaz nodded. "Have you ever told Wren that her eyes are pretty?"

"What? No." Graham could feel the back of his neck burning.

"Why not?" Tal asked, with dancing eyes. "Don't you think her eyes are pretty?"

He glared at his friend and shook his head. "Come on. This isn't about me."

"Wren does have pretty eyes," Jaaz said. "You should tell her before someone else does."

Tal grinned. "I agree."

"Solia could come out any minute. Do you want to waste whatever time we have talking about me?"

"Right, right," Tal said. "Graham is right. Don't tell her what color her clothes are."

"Should I bring up the duck thing again?"

"Let's not do that one again," Tal said. "We don't want to overdo it. Focus on her eyes or smile this time."

Graham winced. "Eh ... maybe you should compliment her on something besides how she looks."

"What do you mean?" Jaaz asked.

"Like something about her personality. How good she is at something. Something that's not shallow."

"Okay ... like what?"

"Isn't there something you like about her?"

Jaaz nodded. "Sure. I've liked her ever since she punched me in the nose."

"Why?"

"I don't know. Can't you guys tell me what to say?"

Graham ran his hand through his hair. "I don't actually know her. Tal's probably the one who knows her best."

Tal blew out a slow breath. "Good things about Solia, good things about Solia," he muttered. "Hm. Most of her personality traits aren't things I would point out to her. Good things about Solia. Well, she's determined. She usually gets what she's after."

"Yeah, like when she wanted my pencil," Jaaz said. "I wouldn't have given it to her if she hadn't punched me."

"It won't matter what you say," Tal assured him. "She'll take one look at Zebra and be absolutely enthralled."

"I still can't believe you named the zebra Zebra. That isn't very creative," Graham said.

Tal ran his hand over Zebra's mane. "Since she's the only zebra here, I think it's fitting."

Graham searched around in his pocket and pulled out a vial. "Everyone take a sip of this," he said, handing it to Tal. "I'm hoping it helps us not get sick if we have to send messages."

Tal took a swallow and handed it to Jaaz. "Wow, that tasted a lot better than last time."

"It's because I used some berries from the maze. I wish we all had filled our packs with them. I don't think Vork is going to be accommodating if I go searching for more."

"The door's opening," Jaaz said, in a high-pitched squeak.

"Okay," Graham said, ducking behind a large tree. "Remember what we told you." Tal laid on his stomach behind the foliage and rested his chin on his hands.

Jaaz walked onto the paved sidewalk that went to Solia's house with Zebra. Graham peeked out, making sure he wouldn't be seen. Solia walked quickly down the walk. Her hands were clenched, and she was almost stomping. She paused when she saw Jaaz and Zebra. She looked Zebra up and down. Her wide eyes were red. Graham wondered if she'd been crying.

"What are you doing here?" she asked, still studying the animal.

"I thought you might want to see Zebra. Isn't she neat?"

"What is it?" she asked, reaching out and touching Zebra's nose.

"I'm not sure. She's my friend's pet. I guess she came from another world," Jaaz explained, puffing out his chest. "She's pretty friendly."

"It almost looks like a horse. Those stripes are really pretty."

Jaaz nodded. "Yeah, but that's kind of shallow. She has a great personality. It's not only about how she looks."

"Oh, Jaaz," Tal said, putting his face into his hands. "I'm glad I can't see anything from here."

"Are you saying I'm shallow?" Solia asked.

"Say no and apologize," Graham whispered into Jaaz's mind.

"No. Sorry," Jaaz said. "I'm not so good at saying things."

"It's fine," Solia said, rubbing Zebra's head. "I am kind of shallow."

"Whoa," Tal said. "I never would have thought she would admit that."

"That's okay," Jaaz assured her. "Everyone is shallow sometimes. Like I would tell you that your eyes are ridiculously pretty, but instead I'll tell you that you're determined."

"Um, thanks?" Solia said, studying Jaaz with her eyebrow raised.

"Are you okay? Your eyes are kinda red."

Solia clenched her jaw. "I'm fine. I had a fight with my dad."

"I'm sorry."

She shrugged. "He can be so annoying sometimes."

"I get that. My dad's annoying almost all the time."

"I needed to take a walk and get away for a while," she said. "Do you want to walk with me?"

"Walk with you?" Jaaz asked, with wide eyes. "Do I?"

*Say yes,* Tal whispered.

"Yes," Jaaz said, turning to the spot Graham and Tal were hiding. "I'll leave Zebra here so my friend can find her," he said loudly.

"Aren't you afraid she'll get lost?"

"Nah. My friend is good at finding her."

"Okay. Let's go this way," she said, pointing. "It's the way I go when I don't want to run into people."

Graham and Tal stayed quiet until the couple disappeared around a bend.

"Well, that went better than I thought it would," Graham said.

"Should we follow them or just hope it goes well?" Tal asked, getting to his feet and walking over to Zebra.

"Let's leave them this time."

"You don't think Jaaz will be awkward?"

Graham smiled and shook his head. "I'm positive he will be. It's another reason I don't want to follow. It's not like we can be in his shadow for the rest of his life. Eventually, it's going to be up to him. We can only fool Solia so much."

"These are good," Tal said, sucking on a frozen berry in Zera's garden. She had an impressive selection of berries. She told Graham he could use anything he wanted. Since everything was frozen, it was hard to know exactly how they tasted, but Graham was having some great ideas for medicine flavors.

"Dryson?" Zera called from the window.

"Yeah Mom?" He still loved calling her mom.

"Jaaz is here. Should I tell him where you are?"

"Sure."

"Taste these," Tal said, handing him a round green berry.

"That's sour," Graham said, scrunching his face. "It has a good flavor, though."

"I've never even seen some of these before. Zera must like her berries."

"Hey guys!" Jaaz said, bouncing into the backyard with a huge smile on his face.

"How did it go?" Graham asked, sucking on something that tasted like a blackberry.

"Way better than we thought it would," Jaaz said, grinning. "She told me all sorts of stuff that's been bothering her lately. I don't believe she's as bad as you guys think she is. She just needs a good friend to talk to. She's so busy trying to be popular and perfect that she doesn't have anyone to tell her problems to."

"I guess I can understand that," Tal said, tossing a berry into the air and catching it in his mouth. "I'm not joining team Solia anytime soon."

Jaaz beamed. "She said I'm a superb listener, and guess what else? When we walked back to her house, she kissed my cheek! Can you believe that?"

Tal grinned and chuckled. "It's hard to believe, but sure."

"It was great. She said we can be secret friends."

Graham narrowed his eyes. "Secret friends?"

"Yeah," Jaaz said. "She said we can be friends when nobody is around, but not at school or anything. I have to pretend I don't know her at school."

Tal smirked. "Now that I believe."

As soon as Wren came to the end of the slide, she sprang to her feet and said, "Light." The cave lit up, and she grabbed a few handfuls of yellow dust from the floor and stuffed it in her cloak pocket. She hoped it was enough. She also hoped she had made the right decision to send Zalliah on the other slide.

She sprinted down the cave hallway as fast as she could. Hopefully, Zalliah would take her time. Whether Zalliah was bad or good, she had come into the cave after Wren told her she couldn't. She hadn't accidentally stepped in.

"Padmire!" she called. "Can you hear me?" She kept running and didn't wait for an answer. Despite running, she was freezing. She was still wet from the rain at Boztoll. Sen had taught them a way to heat themselves from within. He was good at it. That was how he pushed out the heat the last time they were here. She could make herself warmer, but she hadn't pushed any out yet. Running this fast was preventing her from doing any of it.

She came to the pit and grabbed the rope. Tal wasn't here to catch her, but she had gotten better since then. Wren swung across the pit and dropped off on the other side. She almost smiled when she landed on her feet. She had purposely sent Zalliah on the other path because she would have to climb the gigantic wall. This way was quicker and slightly easier.

Wren paused for a moment and listened. All she could hear was the water ahead. "Padmire, if you can hear me, I think the woman on the other path is evil. I'm not positive, but I have a strong feeling about it." She took a shaky breath and ran towards the water. This would be her first time crossing by herself. She could see no reason to wait and talk herself out of entering the cold murky water. As soon as she reached it, she stepped in and waded as quickly as her legs would move, trudging through the chilling river.

When she stepped out on the other side, she gave herself a few moments to warm up. Being too cold would only slow her down. With luck, Zalliah would be fascinated with the cave and would go slowly and carefully. She concentrated on the heat and felt her clothing slowly dry.

As soon as she stopped shivering, she ran. It didn't take her long to get into the final cavern. She glanced around at the silvery walls and the colorful bubbles. She grabbed a red bubble and squeezed it. The bubble was thick and solid. She shook it and watched the liquid inside swish around.

She couldn't let Zalliah get out, not until she knew if she was innocent or not. She threw the bubble as hard as she could at the platform and watched it shatter and something red splattered all over the rhyme that told a person how to get out of the cave. It looked like paint. She squatted and used her hand to spread the paint so that none of the words were visible.

"Well, that's a lovely mess," Padmire said from behind her.

Wren spun around. "Padmire! I hoped you would come."

"I could have caught up with you earlier, but you were in even more of a hurry than usual. Remember my slow feet?" He asked, holding up his webbed foot.

"Sorry. I wanted to get through before Zalliah."

"And who is she?"

"She was one of my teachers. She was the one who sent us here for the first time. I thought she was trying to help us, but now I believe she was trying to get into the cave."

"She was most definitely trying to get into the cave," Padmire said. "I can sense a person's intent when they come here. I can also sense what they need. The need I sense in her is completely self-serving. While I was trying to catch up to you, I was trying to come up with something fitting to gift her with."

"I wish you didn't have to give her anything," Wren said, crossing her arms and resisting the urge to yawn.

"It isn't horrible to give something to a greedy person the first few times they come in. There are plenty of useless things to give them. The problem comes when they enter so many times I run out of nonsense and have to give them something better. Anything in particular you want for yourself?"

"Like magic?"

"Yes."

"Stronger levitating would be nice. I could have used it a few times."

"Done."

"I need to leave before she gets here. If she's stuck, then I can get help."

"The woman is coming slowly because she waited at the bottom of the slide for you for a while. She yelled for you a few times and finally muttered something about teenagers and was on her way. She doesn't appear to be in a hurry."

"Should I go without her or wait? I can't decide. I should send a message to the others and have people waiting for her up top. Sen is the only one that can teleport, so I might need to go up to make portals."

"Who will you send for?" Padmire asked, throwing his skinny arms in the air. "Your three little friends or someone more useful? The giant could be helpful."

"I'll probably send for Graham and Tal. Ming Li wasn't doing too well the last time I saw her. She needs to rest. I'll send for Brake. I don't know about Brog. I'm not sure if he's one of the giants that drank the ellebasi serum or not."

"Well, acting now is probably your best bet. The woman is coming faster. At the speed she is going, she will reach us in less than five minutes."

Wren nodded and pictured Graham in her mind. She closed her eyes and whispered, *"In the cave with Zalliah."* She waited and tried to release any tension in her mind that might give her a headache.

*"Okay?"* Graham's voice said in her head.

*"Zalliah might be Dark Cloud."*

*"WHAT!"* Graham's voice boomed through her head. It felt like it was ricocheting off her brain. *"Sorry,"* he said at normal volume.

*"They drugged the giants. Bring help."*

*"Done."*

Padmire scratched his head. "That is such a strange thing humans do. Inefficient to be sure. I don't know why you would need to do that with your friends now that you have all exited the cave together."

"It's the only way we can communicate."

"It's not. Once the five come out of the cave, they are bound in certain ways. All you need to do is put your fingers to the side of your forehead and think about the message you want to send. No one can hear you and it doesn't have side effects. No headaches, no sick stomachs."

"That would have been nice to know earlier. Can we do it with anyone?"

"No. Just the five of you. Others used to be able to do it, but those days have passed. The woman is almost here. She's coming faster."

Wren shifted her weight from one leg to the other. "Should I stay?"

"If you don't, she might get away."

"Alright," Wren said, looking towards the entrance. "But what if–" Padmire was gone. "Lovely," she muttered, rubbing her cold arms. How was she going to stall until the others came? She put her fingers on her head and thought about Graham. *Tell me when you arrive,* she thought.

"Ah, there you are," Zalliah said, entering the cave. She was dry, so she either didn't go into the river or she knew Sen's trick. "Why didn't you come after me? I thought that was what we agreed on."

"Sorry. After you slid down, I realized the other way was easier for me."

"Nice to let me know," she said, raising her eyebrow. "This room is incredible," she said, taking in her surroundings. "Is this the magic?" she asked, reaching out for a blue bubble.

"People take a bubble out with them and they say it absorbs into you on the way out," Wren said, not wanting to lie outright.

"Fascinating." she said, shaking the bubble and holding it to her eye. "I've studied this place for a good portion of my life and now that I'm here, I am realizing how much we didn't know. My studies made it seem like it would be a lot harder. I'm a little surprised you came through before me. I mean, I waited for you, but I still would have expected it to take you longer."

"This is my third time through. I figured there wasn't a point in lingering."

"So what do the colors mean?" she asked, grabbing a red bubble with her left hand.

Wren shrugged. She wasn't sure how much she should tell. Zalliah's eyes were sparkling like a kid at a toy store. She grabbed another bubble and held all three in one hand. Wren hoped Graham would hurry. She wasn't skilled at stalling.

"We should take one of each, just to be sure." Zalliah grinned, plucking another bubble out of the air.

Wren rubbed the back of her neck. "I think you can only take one."

"You think? Have you tried?"

"No."

"Well, I'm going to try. I am not privileged as you are to come and go as I please."

"What if something bad happens?"

"Such as?" she asked, as she continued grabbing bubbles.

"Since the cave closed because people kept coming in and getting more powerful, it seems like that would mean they couldn't get it all at once."

"Oh Wren. You see things much differently than I," she smiled. "Have you ever thought that people assumed they had to come back? Perhaps nobody ever tried to take it all at once."

"Why would you want all the power?"

"I don't necessarily want all the power, but if it's here to take, then we might as well."

Wren's tired eyes burned as she watched Zalliah grab as many bubbles as she could fit in her arms. Why was it so hard to know who to trust? Until today, she had believed Zalliah to be one of their supporters, but the more they talked, the more she was ready to believe it was the complete opposite.

"What's this?" Zalliah asked, looking at the red mess Wren had made earlier. She rubbed her foot against the drying liquid and tried to read the smeared writing. "This is fresh paint. Did you do this?"

Wren hid her red fingers behind her back. "One bubble broke."

"What did it say?"

"Nothing important."

"I've been teaching teenagers for years. I know when I've been lied to," she said, locking eyes with Wren. "What are you hiding, and why?"

Wren narrowed her eyes at her teacher and took another step back. "I find it hard to trust someone who came in here when I asked them not to."

"Oh please," Zalliah laughed. "Anyone would have come in. This is a place of legends."

"I know plenty of people that wouldn't have come in," Wren told her.

"So what do I need to do to make you trust me?" Zalliah asked, still scanning the cavern for any color bubble she might have missed.

"Do you need my trust?" Wren asked. "You already have what you want."

"I've always liked you, Wren," Zalliah said, shifting her arms so that she wouldn't drop anything. "You were always quiet, but persistent. I watched you fail over and over, but you kept trying even when I could tell you hated it. I can tell you've gotten stronger since I last saw you. You must have made good time getting through the cave today. It was impressive."

"Yes, I'm much faster than the last time I came here. You all must have been waiting a long time for us to get out the first time."

Zalliah lifted her chin and grinned. "You think you have this all figured out, do you? So what do you know?"

Wren shifted her feet and kept her gaze on Zalliah. "I know you're here for power, and that is enough to make you untrustworthy. It's pretty obvious you're not the person you want us to think you are. I believe you're one of Gorbin's followers and a member of The Dark Cloud."

Zalliah's smile slid from her face. "One of Gorbin's followers? Don't make me vomit. Gorbin is nothing but a sheep that does whatever I tell him to do. If Gorbin were left to himself, he would self-destruct in a matter of days. He needs to be instructed in all things. I let Gorbin do the talking because he gives off a large, intimidating presence, but that is all he's good for."

"You don't seem to deny anything about The Dark Cloud." Wren said, tightening her fists.

"Well, I don't see a point. You seem to be convinced already. It's good to have someone finally see that I deserve credit for so many things. Of course, that means I can't let you leave this cavern alive, and that is unfortunate, because I do like you."

Wren knew she should feel panicked at this point, but she just felt irritated. How had they let the enemy train them?

"You look disappointed. Is there anything you want to talk about?"

"You want me to tell you my problems before you kill me?"

"If it will make you feel better."

"I can't believe you are part of The Dark Cloud. It's kind of sad."

She laughed. "Part of The Dark Cloud? I am The Dark Cloud. If it wasn't for me, the entire organization wouldn't exist. Things could go better for you if you joined of your own accord. I want you to see my vision. We aren't so different in what we desire."

Wren raised her eyebrow. "Not that different?"

"I can see explaining it to you would be a waste of time."

"Why are you a gym teacher? That seems like a strange alias."

"I love activity. Young people don't realize the benefits of exercise. I want to inspire people."

"You are really confusing."

Her eyes sparkled. "Yes, and I like it that way. Of course, helping young people isn't my only motivation. I've seen enough auras in my time to know that the five who would enter Meegore would be young. I assumed they would be older than you, but the school seemed as good of a place as any to keep watch."

"Lucky you chose the right school," Wren said, wondering how long she could stall.

"Yes, it was. I had some clues. I knew if Brake ever found his son, he would send him to school there. Years ago, I saw an aura around Brake's baby that led me to believe his son would be important. When I saw Graham that first day, I saw the brightest aura I've ever seen around the four of you. It was the best day of my life. I knew Graham had to be Brake's son."

"Were you working with Governor Briggs?"

"No, why?"

"Governor Briggs has a weird interest in Graham."

"Of course he does. I told him years ago that Brake's son would be important. As soon as he heard Brake had a nephew here, he probably came to the same conclusion as I did."

"Is Briggs part of your group?"

"Goodness no," she sneered. "That man is almost completely useless these days. I'm shocked every day that a man like him could have a son like Tal. I hope he appreciates that boy. He's the only good thing I've ever seen come from Briggs."

"What are you doing at Boztoll?" Wren demanded.

"Doing?" she asked, pushing a lock of brown hair behind her ear.

"What's the point of sending non-magical people to Boztoll and then making them miserable?"

"Ah," Zalliah smiled with a gleam in her eye. "That is my brilliant plan. I'm not actually going to kill you, you know that Wren? You are valuable to me. People will do almost anything when they feel they owe someone. The people in Boztoll are miserable. Other non-magical people around the world are afraid."

"Yes, and I don't see why you would do that, even if you are against them."

"Oh, I'm not against them. They are all part of my plan. The people are not enjoying their lives at this time, but think of how happy they would be if someone were to fix their situation? I'm not against the non-magic. I'm against the rest of the world."

"I'm lost," Wren said, narrowing her eyes. Perhaps Ming Li was right, and Zalliah was crazy.

"The Dark Cloud is full of dangerous people. Dangerous and stupid people. My plan is to save all the non-magical people by destroying The Dark Cloud. After that, they will be indebted to me. The sun will shine on Boztoll again, and if that isn't enough … you will take them all to Meegore to get magic."

"No way!" Wren said, making a fist.

"I love the fire in your eyes. It doesn't worry me though. You'll be easy enough to control. Once all the non-magic have magic, I will turn them against everyone else. It won't be hard. What has this world done for them?"

"There are way more magic people than non-magic."

"Yes, and I have a plan, to be sure. My plan has been in place since before your father and his silly friends tried to defeat us when you were a babe. I must be doing something right. They never suspected me in all these years. They were fighting my organization before they even realized what it was and all they managed was to take out a few weak members. Don't worry, I'm not greedy. I'm not trying to take over the world. Just Akkron."

"So you are going to turn on your own followers?"

"Not all of them. Just the ones who aren't useful. You would be amazed at how many people will join a cause they understand very little about. Look at the girl Melly. We never let her in our group and she kept trying. Why? She didn't even know. People want a place. I have enough disposable people to make it very convincing when I destroy them. It won't be hard to get elected governor after that."

"I'm never going to help you."

Zalliah winked. "You will. Don't worry, you will have a central part in it all."

"Worry? What's stopping me from telling your followers your plan?"

"Ah. You don't know which ones are already part of it. I'll let the more valuable members know they are to kill you immediately if you tell them anything. Do you want to risk telling the wrong person? I don't want to kill you or any of your friends. It's a lot more convenient to have five of you in case something happens. That doesn't make you indispensable though, and I will eliminate you if I must."

"I'm still not going to help you," Wren said, glaring at Zalliah.

"Oh, you will, and for the same reason you will tell no one about this conversation. You don't want any of your friends to get hurt. You realize you've surrounded yourself with a lot of people. I won't start with your four friends of course. There are lots of others I can get rid of before them and I will start with your father. I'm not planning on letting you go anyway, so there is little chance of you telling anyone."

"Ahhhhhhhh!" Padmire's shrill voice yelled as he dropped from the ceiling and landed on Zalliah's head. She dropped all the bubbles as she screamed and tried to pull him off. The bubbles shattered and left colorful splatters on the floor.

"Get off of me!" she yelled, as Padmire bit her head.

"Leave Wren!" Padmire called. "I have this under control. Take the book this time."

Wren sprinted to the stand that held the old book and grabbed it. She started for the platform, but paused when Zalliah threw Padmire to the floor.

"You will regret that you little monster!" Zalliah said, smoothing her hair back into place. She kicked Padmire, and he went flying across the room. Wren sprinted towards Zalliah and smacked her in the head as hard as she could with the book, causing her to fall over.

"Not with the book!" Padmire scolded as he hopped to his feet. "I told you I have it under control."

Zalliah grabbed Wren's ankle and pulled, knocking her to the floor. Wren dropped the book as she fell and Zalliah scooped it up and jumped to her feet.

"The Silver Eclipse?" she said, looking smugly down at Wren. "We can't have that now can we?" She held up her hand and fire sprung onto her palm. She held the book over it and as the spine caught fire, Padmire jumped on her leg and gnawed on it. She screamed and dropped the smoldering book next to Wren.

Wren grabbed the burning remains and tried to run towards the river. It was burning too fast, and she had to drop it.

"Get out of here!" Padmire roared. Zalliah was trying to pry him from her leg.

"I'm not going to leave you!" Wren said, sticking her burned finger in her mouth.

"She can't hurt me. I'm a bungle," he smiled. He snapped his fingers and the flames on the book went out. He snapped again and Zalliah fell to the ground. Her arms and legs stuck to her sides. "Now go."

"Why didn't you do that to begin with?" Wren huffed as she jogged towards the platform.

He shrugged his blue shoulders. "It's not very satisfying."

Wren jumped on the platform and glanced back. Zalliah was struggling with her invisible bands and yelling something she couldn't understand. That was the last thing Wren saw as she swirled away, back to the island.

# CHAPTER 17

Graham dropped to his hands and knees as a huge rock flew past his head. This was getting crazy. As soon as Sen teleported Graham and Tal to Meegore, they were ambushed by Dark Cloud members in black cloaks and masks. He couldn't tell how many there were, but it was significantly more than last time. This was beginning to feel too familiar. Graham had sent a message to Sen as soon as Wren contacted him. Sen met them and teleported them all to the island. He left almost immediately to get Brake and any other reinforcements he could find.

A few giants had joined the fight, wielding massive clubs at the attackers. The only giants with swords were Mot and Tnarg, and so far, Graham hadn't seen them. While Tal used his powers to trip the masked assailants with tree roots, Graham defended himself by shooting ice until he was knocked to the ground. He quickly regained his footing and turned to confront a large masked man throwing rocks.

The man laughed as he took a step towards Graham. Graham's arms flew to his sides and he couldn't move them. It was definitely Gorbin.

"You thought you could swoop in and take Fria and there would be no consequences?" he growled.

"Don't try to pretend this is about vengeance," Graham spat, as he struggled to free his arms. "You attacked us twice before we captured Fria!"

Gorbin slapped him hard across the face, causing him to flinch in pain. Whatever invisible force Gorbin was using to keep his hands down didn't let him fall. Gorbin hadn't held back, and the whole left side of Graham's face throbbed.

"It was about power, but now it's about vengeance," Gorbin said, poking Graham in the chest with his finger.

"Ahhhh!" Gorbin yelled, as Zera jumped on his back. Graham was stunned as he watched her expertly take down Grobin. It all happened so fast, Graham wasn't even sure what she did. One second she was on his back, the next he was on the ground with her boot pressed into his back. Graham's arms were immediately free of his invisible hold.

"You never touch my son!" Zera said, grinding her heel and stepping off. She grabbed Graham in a quick hug. "Are you okay?" she asked.

"Yeah, thanks Mom."

"I have this one," she said, pointing at Gorbin. "Go help Tal."

Graham scanned the crowd of fighting people and saw Tal. He was sinking into the ground and the dirt was to his knees. A bunch of birds were flying around him, squawking at anyone who came near. A figure was pointing their hands at him and Tal was slowly getting deeper into the earth. Tal was ignoring the figure completely as he wound his hands around each other and caused roots to burst from the ground and whip at the villains fighting the giants.

Graham hurried toward Tal and the figure, wondering why Tal seemed oblivious to the person making him sink. He readied himself to shoot ice at the figure when suddenly, a bolt of lightning struck the ground in front of the villain, blowing earth twenty feet into the air. Graham shielded his face with his cloak and coughed as the dust and debris settled. When the air cleared, the figure was lying on the ground. Tal was still coughing from where he was stuck in the earth, and Wren sped towards him.

"Why are you just standing there?" she asked when she reached them.

"I had him under control," Tal said, shaking dust from his hair. "He seemed so pleased with himself. I figured I would let him keep going, so he wasn't bothering anyone else."

"Under control?" Wren asked, putting her hands on her hips. "You are stuck past your knees."

Tal kept flinging roots and vines, but he managed a crooked smile. "Am I though?" He raised his hands into the air and was lifted through his dirt prison. "I was standing on a big root. I could have gotten out at any time."

Wren sighed and looked around. "I saw my dad, Austra, and Brake. They seem to be doing fine. Who needs help?"

Graham glanced around. Everyone was busy fighting, and nobody was paying attention to them. The Dark Cloud must see them as less of a threat than the adults.

"Don't do the lightning thing unless you have to," Graham said. "I think I'm going to have dust in my lungs forever."

"My ears are ringing," Tal said. "Can't you make quiet lightning?" he asked, tripping someone who was fighting with Austra.

"I doubt it," Wren said.

"So, is Zalliah really Dark Cloud?" Graham asked as he scanned the trees.

"Yes," Wren said, clenching her teeth. "According to her, she's the one who started it all."

Tal pointed toward the sky and a flock of birds appeared, swooping down to peck at the masked figure's heads. The Silver Eclipse and the giants took advantage of the distraction. Tal made roots pop out of the ground and cover anyone who fell in makeshift cages.

"Silly boy," Zalliah said, walking towards them, waving her arms. The root cages became untangled and pulled back into the ground.

"Oh, no you don't," Tal said, flinging a vine at her. She raised her hand, and the vine dropped harmlessly to the ground.

"Come now. You don't believe you can best me at something I've been doing since before you were born?"

Graham raised his hands and shot as much water as he could manage, knocking Zalliah to the dirt. He kept the water coming until his arms ached.

"We should have known she could control plants," Tal said, as Zalliah pushed herself onto her hands. Her wet hair dripped onto the muddy mess beneath her. "She made that gigantic tree appear in the gym for our training, remember?"

"Yes, you have all been a little slow piecing things together," she said, wringing out her hair as she got to her feet.

"Zalliah! We have to go!" Gorbin called, running up to her. His mask was gone and his head was bleeding. He also had a slight limp.

"I say when it's time to go," she snapped.

"Look around! We are losing!" Gorbin said, extending his arms. "The giants are waking up and joining in."

Zalliah glanced around at all the fallen members of her group and glared at Graham. "Don't leave anyone," she said to Gorbin, and she disappeared. Gorbin nodded and ran, calling out orders.

"What do we do?" Tal asked next to Graham. "Do we let them go?" Members of The Dark Cloud were slowly disappearing and taking their fallen members with them.

"I don't know," Graham said, looking at Wren. She shrugged. Brog had two limp figures under his arms. They weren't going anywhere. All the Dark Cloud that was still conscious were gone.

"Are you all okay?" Zera asked, joining them. They all nodded as Sen and Brog approached. Brog dropped the two people he was holding and five more giants came up, each dropping one more unconscious figure. It looked like The Dark Cloud was leaving members behind. Zalliah wouldn't be pleased.

"What are we going to do with them?" Brog asked.

"We will lock them up," Zera said.

Brog nodded. "I hope your prison is big enough."

"It is," Brake affirmed, walking towards them. "We made sure it would be bigger than we could ever need." He patted Graham on the back and smiled at Wren.

"We will bind them," Brog said, nodding at the other giants. "One of them is dead." Wren shivered.

"We can't let something like this happen again," Drew said, pulling Wren into a hug. "How did they all end up here, anyway?"

Wren stared at her boots. "Ming Li and I were in Boztoll and The Dark Cloud started causing problems. After we took care of it, Zalliah grabbed me and brought me here. She told me I needed the yellow dust inside the cave to cure the giants. They were all asleep from the ellebasi serum. I went into the cave and she followed me."

"So it was a setup," Brake said, rubbing his chin. "They probably didn't realize you could call for help. Ellebasi wears off like ailam powder. You don't need anything to wake them."

"Who is that on the ground over there?" Sen asked, pointing. Graham looked over and saw someone lying in the dirt. He couldn't tell who it was from this distance.

"Austra!" Drew yelled, running in her direction. They all chased after him, but he was going fast. Graham didn't know he could run like that. All the adult members of The

Silver Eclipse kept surprising him. When Drew reached her, he dropped to his knees and bent over her. "She's breathing!" he said, rolling her onto her back.

Graham kneeled on the other side of Austra to see what was wrong. He pulled out all of his vials as he watched her. The side of her face was cut and there was blood on her side. Her eyes fluttered, but didn't open.

"Austra? Can you hear me?" Drew asked, gently shaking her shoulder. Graham poured some cream into his hand and rubbed it on Austra's face. The cut immediately started to fade.

"What happened to her side?" Graham asked, scratching his head. Sometimes knowing how to help people was hard. It looked like she was burned.

"That's what one guy I hit with a fireball looked like," said Sen.

"Okay, a burn," Graham mumbled as he searched through his bottles. "I know I have something. Burns are painful, that's probably why she's still unconscious."

"Dad, are you okay?" Wren asked as she approached him. Graham peered up to see tears running down Drew's face. His eyes were rimmed with red and his lips were pulled down in a thin line. Graham couldn't remember ever seeing him so distraught.

"Just focus on her," Drew ordered roughly, when he caught Graham observing him.

"This should help with the burn," he said, pouring a green liquid on the wound. It bubbled and made a sickly gurgling sound. Graham cringed at the sight and smell, as did everyone else. He took another bottle and dumped in on top, clearing away the green mess. When it was gone, the skin was as good as new.

Austra breathed deeply, and her eyes slowly opened. "What happened?" she asked, sitting up with only a little difficulty.

"You're better!" Drew beamed, wrapping her in a relieved hug.

"Yes, I'm fine," she said, patting him awkwardly on the back. Drew stood and pulled her to her feet.

"Are you alright? Are you weak?" he asked.

"No, I feel completely fine. What happened to my dress?" she asked, running a hand over the scorched red fabric.

"I thought you were dead," Drew said, wiping his eyes. "That was the worst feeling I've ever felt."

Wren's eyebrow raised as she watched her dad.

Austra tilted her head and gazed at Drew. "I'm fine."

He wrapped his arms around Austra's waist and smiled as her eyes widened. "Good," he said, as he leaned in and kissed her. Graham peeked at Wren. Her eyes looked like they were about to pop out of her head. He couldn't tell if she was upset or shocked. Brake had his arm around Zera and they were both smiling as they walked slowly away. Sen was focusing on the ground and Tal had the biggest grin Graham had ever seen.

Graham kept waiting for Austra to punch Drew, but she wrapped her arms around his neck and kissed him back. Graham turned and caught up to his parents. Zera's eyes lit up as she linked her arm through his.

"That was unexpected," he said. "Not to mention uncomfortable."

Zera grinned. "Unexpected, but overdue."

"That's for sure," Brake agreed.

"Overdue?" Graham asked, wrinkling his nose.

"Yes," Zera nodded. "Drew and Austra have been in love for years. He told her when they met he had an arranged marriage, but they still fell in love. Drew tried his best with Magnalee, but they were so different."

"I thought Austra hated him," Graham said, shaking his head. "She always acts like he's so goofy."

"I think it was her defense," Zera said. "If she could convince herself, she wouldn't get hurt again."

"Oh, my heck!" Wren said, catching up to them. Tal and Sen were right behind her. "How long is a kiss supposed to be? This is getting ridiculous," she muttered, peeking over her shoulder.

"It's been a long time coming," Brake laughed. "They are making up for lost time."

"Lost time? What are you talking about? I don't know why she's kissing him. She's always rude to him."

"It's probably time for you to have a talk about the past with your dad," Zera said.

"You mean they've done this before?" she squealed.

"Not that I know of, but there is a story there."

Brake threw his head back and laughed. "Hamble is going to be so mad he missed this."

"He really is," Zera agreed, her lip twitching.

"What magic do you think Zalliah got in the cave?" Graham asked, changing the subject for Wren's sake.

"I can tell you that," Padmire's shrill voice said from a nearby tree. They all turned and stared at the bungle. "This was one of my better ideas," he smiled, rubbing his hands together.

"So what is it?" Tal asked.

"She now has the power to secrete blue mucus when she's in the dark. And by saying she has the power, I mean she has no choice. That mucus will glow in the dark. I hope she gets a headache tonight when she tries to sleep, because it's not a gentle glow. It's a bright, hurt your eyes kind of glow. No more slinking in the shadows for her!"

"That's the best thing I've heard all day," Tal laughed. "Remind me to never get on your bad side, Padmire."

"Only happy to help," he said, showing his pointed teeth. "Giving her that magic was almost as satisfying as chewing on her leg."

Tal laughed again and smacked his knee. "You chewed on her leg?"

"And her head."

"I wish I had seen that."

"Wren did. She can confirm it."

Wren gave a half smile. "It was something I won't forget in a hurry," she said, taking one more glance behind her.

"So what's going on with Ming Li?" Brake asked.

Wren quickly filled them in on what they had missed in Boztoll. "I wonder why they would plan something in Boztoll and Meegore on the same day."

"I'm sure it's all connected," Brake said. They all stopped and watched the giants finish tying up the final Dark Cloud members.

"Ming Li is completely worn out," Sen said. "I couldn't get her to wake up, but she seemed to breathe fine. Some of us should probably go back and check on her."

"I'll come," Tal said. "I want to see Boztoll when I'm not tied up."

Wren yawned. "I should go check on her, too."

"I am sure she is fine," Zera said, looking Wren up and down. "You should go get some sleep. You look like you are about to fall over."

"I am tired," she admitted.

Graham shifted to his left foot. "Do you think I can help her? Is she hurt at all, or just tired? I don't think I can help tired."

"She's only tired," Sen said.

"Alright," Zera nodded. "Sen can take Tal, and I'll go to make sure she is okay. Wren can make a portal to the prison and Brake, Drew, and Austra can help take the prisoners. Hopefully, Wren has enough energy to open another portal so she and Graham can go rest at the meetinghouse."

Wren tilted her head. "I think I can."

"I don't need to rest," Graham protested. "I've spent the last few days playing matchmaker with Solia and Jaaz. It's super weird, but not tiring."

"It would be better for you to go back anyway," Zera said, motioning with her head towards Wren. Graham looked over to see Wren with her arms crossed, frowning at the place Drew and Austra were still embracing.

"Alright," he agreed. "Let us know if you need us."

How was it possible to be this tired and not be able to sleep? Wren rolled over in her bed at the meetinghouse and tried to get comfortable. She had slept for about an hour and now all she could do was roll restlessly around. Her body felt drained of any energy, and at the same time, refused to relax.

"I give up," she muttered, sitting up and throwing her pillow at the wall. She climbed out of bed and trudged over to the kitchen. When she entered, she found Graham and Hamble relaxing at the table, eating sandwiches and drinking hot chocolate. They both glanced up. Graham frowned and Hamble's eyes twinkled.

"I take it you didn't get any rest?" Hamble asked, getting off his chair and going to the stove. Wren sat at the table and glowered at it. "Graham was filling me in on everything I missed. I'm sorry I wasn't there to help." He brought a mug of hot chocolate over and set it in front of her.

"Thanks," she muttered, taking a sip.

"You should have slept more," Graham told her.

"I know. I kept tossing and turning."

"That's what your hair would suggest," Hamble chortled.

Wren tried to run her fingers through her hair, but there were too many snarls. She growled under her breath and took another sip.

"I wish I could have snarls like that," Hamble said, rubbing his bald head. "Ah well, we all get what we get, don't we? Graham was telling me about Zalliah. I'm pretty shocked. I went to school with her. She was always determined, but I never would have expected her to try to take over. It's no wonder you can't sleep. It's probably pretty irritating to find out your teacher was manipulating you."

"I wish that was why I couldn't sleep," she said, staring into her creamy chocolate.

"Oh? What else could be bothering you?"

Graham coughed. "I didn't tell him that part."

"What? What did I miss?" he asked, leaning in. "I can't imagine anything more annoying than Zalliah trying to take over the world."

Wren peeked at Graham and was irritated at the sympathetic look he was giving her.

"I don't want to talk about it," she mumbled.

Hamble looked at Graham, and he shrugged. They all sat in a weird silence for a few moments. Wren felt more angry the more she thought about it. Why was her dad kissing Austra, of all people? Austra was rude and bossy. If her dad wanted a relationship, Wren was okay with it, but not with Austra.

"UGGGH!" she exclaimed, slamming her mug on the table. "Why Austra!" she growled. "Austra is stuck up and thinks she's the only one who knows what's best. I ignore what she's saying most of the time, but I don't have to deal with her that often. What if that changes? What if she ends up in my life?"

Wren stared at Graham. His eyes were wide, and he looked like he wanted to disappear. Why was she over-reacting? They both thought she was crazy. She stole a glance at Hamble, whose eyebrows were knitted together in confusion.

"It might not come to anything," Graham said, putting his hand over hers on the table.

"They were sucking on each other's faces! I've never seen my dad act interested in anyone before."

"Wait ..." Hamble said, his eyes lighting up. "Drew and Austra were kissing? It's about time."

"About time?" Wren asked, wondering if Graham was going to move his hand.

Hamble nodded. "I would say you should ask your dad about it, but we both know he won't tell you anything."

"That's likely," Wren said, pulling her hand out from Graham's and pushing her hair behind her ear.

"When our group first came together, your mom wasn't part of it. Drew knew about his marriage agreement with Magnalee, but he had never met her. Austra and Drew had a lot of assignments together and they became good friends. Austra knew about Magnalee from the start. When she came into the picture, she joined our group. Magnalee fit in perfectly. She quickly became friends with everyone except Drew and Austra.

"I'm not sure why she didn't try harder with Drew, but she seemed irritated by everything he did. This made Austra want nothing to do with her. Austra always saw something in Drew that others didn't. It later came out that Magnalee hadn't known about the contract until after she had fallen in love with her neighbor boy and her father told her she couldn't be with him.

"Austra begged Drew to run away with her to the other continent, but he wouldn't go against his parent's wishes. Austra was actually quite pleasant until that. Well, pleasant if you didn't get in her way. After Drew and Magnalee were married, Austra became distant. It's gotten worse over the years. We all thought things would work out after Magnalee died, but Austra had already built a barrier over her heart."

"She should leave it there," Wren sulked.

"Put yourself in her place," Hamble suggested. "Imagine you were to fall in love with Graham and then found out he was going to marry Solia. How would that feel?"

Wren sat up straight and tried to ignore the heat creeping up the back of her neck. Hamble had as much tact as Ming Li. She didn't risk looking at Graham.

"I'm going to go think," Wren said, jumping up and walking to the meeting room. Once there, she paced back and forth, trying to decide how to feel. She could put herself in Austra's place easier than she would like to admit. The meeting room was so boring. The white walls could use some pictures. She sank into a red chair and let out a slow breath.

"Hey," Graham said, poking his head through the door. "Can I come in?"

"Sure."

"Are you okay?" he asked, sitting in the chair next to her and putting his arm around her shoulders. She stiffened.

"Probably. It's just so weird."

"It is a little weird."

"What if they get married? That would make Austra my step-mom." She made a face.

"Hamble's right though," Graham said, quietly. "Think about how you would feel in your dad or Austra's place."

"I know," she said, looking at her hands. "I know I'm wrong, but it's still hard."

"That's okay. Sometimes we need time to come to terms with things. Think of how happy your dad could be. Also, that might have been the end of it, so you might be worried for no reason."

"That's true," Wren said, rubbing her temples. "It's not like I'm ever home these days anyway. It might not even affect me. I want my dad to be happy. I don't know. It all feels confusing."

A clicking sounded and Drew fell into the room. He bounced off the large cushion and landed on his feet.

He narrowed his eyes at Wren. "I thought you were resting. You look worse than before."

Wren pursed her lips to hold back whatever wanted to come out of her mouth.

"What's going on?" he asked, glancing from Graham to Wren. Graham quickly pulled his arm onto his lap.

"Nothing," Wren said, folding her arms.

He raised one eyebrow. "That didn't look like nothing."

"Are you serious?" Wren growled. "We were sitting here talking."

"Talking?" Drew said, rapidly tapping his foot against the floor. "It appeared you were in some sort of hugging position, and you aren't old enough to be hugging boys."

Wren rolled her eyes. Why was she so mad? "A hugging position? That was not a hugging position. Graham was trying to make me feel better about the fact that my dad was sharing spit with Austra."

"Oh," Drew said, rubbing the back of his neck. "You saw that?"

"Saw that?" Wren said, standing up. "Everyone saw that! I didn't think it was ever going to end. It's burned into my permanent memory. You are way too old to be kissing people!"

"How old do you believe I am?" he asked, throwing up his arms. "I am not as old as your imagination thinks. Lots of people don't even get married until they are my age!"

"I don't get why you can get mad at me because Graham kind of has his arm on me when you're out kissing Austra. It's so gross!"

"I'm old enough to be kissing someone. You aren't."

"I'm not kissing anyone!" Wren peeked at Graham. He was slumped in his chair, most likely wishing he could disappear into the cushions. "I know I'm wrong here, okay? It was weird and I'm so tired and it will all be fine tomorrow." She turned and ran from the room. She hoped it would be fine tomorrow.

Graham ran a hand over his face and wished he could disappear.

"She's a little uptight today," Drew said, sitting on a chair across from Graham. Great, he was going to stay.

Didn't he have the decency to leave and let the tension subside?

Graham shrugged. "She woke up in the middle of the night and she hasn't had time to stop. She probably doesn't even realize what she's saying."

"So, have you two kissed?" he asked, leaning forward.

Graham coughed. "No!" What was wrong with this guy? Did he *like* awkward?

"That seemed like a defensive no," he said, peering closely at Graham.

Graham fidgeted in his chair. At least he could answer honestly. "I've never kissed her. I promise."

Drew rested his elbows on his knees. "Have you ever wanted to?"

"What? What kind of question is that?"

"I'll take that as a yes." Drew watched him squirm, his steely eyes revealing nothing about what was going on behind them. "Wren isn't old enough to date. Sixteen would probably be old enough ... although seventeen might be better."

"I'm not trying to date anyone," Graham said, trying to think of a way to change the subject. There was another clicking sound and Zera suddenly fell into the room. She jumped quickly to the floor and Brake appeared a few seconds later. Graham wasn't sure if this would make things better or worse.

"What's going on?" Brake asked, noting the uncomfortable tension in the room. He looked from Drew to Graham. Graham covered his face with both hands and slid them down so his chin was resting on his hands and his elbows were on his knees.

"I'm just telling Graham that Wren is too young to date," Drew said, shrugging.

"For the last time, I'm not trying to date her!" Graham said, throwing his hands in the air.

"Well, I don't see a reason to put your arm around a girl you aren't planning on dating."

Graham slumped in the chair and stared at the ceiling. "She was upset. I was helping a friend. You're turning this on us because you're trying to take the attention off of you and Austra."

Zera sat by Graham and kissed his cheek. "You should go work on your medicine and your father and I will take care of this. Okay?"

"Thanks," Graham said, hopping up and fleeing the room. Back in the kitchen, he found Wren and Hamble sitting at the table again. "Hey, Hamble. If you don't want to miss anything else, you should probably go into the meeting room."

"Great, thanks!" Hamble said, eagerly making a beeline for the other room.

"Sorry about that," Wren said, looking sheepishly at Graham.

"It's alright," he said, grabbing a banana and sitting. "I think it's a parent's job to embarrass their children."

"Sometimes it feels that way. I hope it didn't get worse after I left."

"It did," Grahams said, peeling the banana. Wren groaned. "Don't worry about it. My parents came in and let me escape."

"I should have reacted better. I want him to be happy."

"Just take some time to cool off. You should try to go back to sleep."

"I feel guilty about too many things. Guilty about my dad, and guilty I'm not checking on Ming Li."

"Ming Li only needs sleep. Do you want a bunch of people peeking in on you when you're that tired?"

"I guess not," she said, fiddling with a napkin. "It was embarrassing when I passed out at that cabin awhile back and everyone was watching me."

"Go sleep and we can check on her tomorrow."

"Alright. And thanks Graham. You're a great friend."

# CHAPTER 18

When Wren entered Valeena's house the next day, Ming Li appeared to be doing better. Hours of sleep must have been what she needed. Her hair, on the other hand, was a disaster. Wren felt like she had been untangling hair forever. Hers was pretty knotted this morning, and now she was spending what felt like an eternity helping Ming Li with hers.

"Ouch!" Ming Li complained from her seat on the floor. "I feel like you're ripping it all out."

"Sorry," Wren muttered from the couch. She had never seen such a knotted mess. It was like a bird's nest.

"Everyone is talking about how long the clouds stayed parted after I used my powers yesterday, but there is no way I can do that regularly. I've never been that exhausted in my life. If I ever do it again, I'm going to pull my hair into a tight braid and wear a shower cap or something."

"If it makes you pass out and sleep for that long, I don't think you should do it again."

"I won't unless it's an emergency or something."

"You missed all the craziness yesterday."

"Ouch! I know. Tal and Sen told me this morning, and I am really sad I missed it. I still can't believe Zalliah is the bad guy. It makes sense when you think about it, but we definitely didn't."

"This is ridiculous," Wren said, wiping her forehead. "You might have to cut it out."

Ming Li grabbed her hair and clutched it over her shoulder. "No way!"

"I'm not going to cut it. I'm just saying I'm not making much progress."

"What if we put something in it?"

"Like what?"

"I don't know. Something slippery. I wish someone here would invent a conditioner."

"I'll try for a few more minutes, and if it doesn't work, we can try something else."

Ming Li sighed and released her hair. "Fine. My head is going to hurt all day. So, are you going to tell me about your dad and Austra? That's even more wild than Zalliah. I thought Tal was joking when he told me."

"Ugh. I've been trying not to think about that," she said as she untangled a few hairs. "It's fine. Austra isn't my favorite person, but I want my dad to be happy, so whatever."

"I kinda thought Austra and Hamble would end up together."

Tal entered the room with his lopsided smile. "Austra and Hamble? They would be the worst couple ever. The snarl is getting bigger."

Ming Li growled, and Wren sighed.

"Let me help," Tal said, reaching out for the brush. "What are those looks for? I have a whole stable full of creatures that occasionally get worse tangles than that. I guarantee my dad isn't the one who gets them out."

Wren stood and relinquished the brush to Tal. He took her seat and started on the mess.

Ming Li cringed and looked at the ceiling. "This is so bizarre. If you had told me a year ago that I would be sitting in Boztoll, with Tal unsnarling my hair, I would have said you were crazy."

Wren smiled and then laughed. She had too much pent up stress. She laughed so hard, tears were running down her face. Tal and Ming Li joined in. A year ago, they wouldn't have been caught dead with Tal. Of course, they had learned that they were misjudging him, and now they were all friends, but they definitely never would have pictured this scene.

Tal hummed as he worked on the tangles. "One time, Lazrelle got a bunch of spiky weeds stuck in her tail and she rolled around in the dirt trying to get them out. It took me two hours to fix that mess. This isn't nearly as bad as that was."

"So, now I'm getting compared to an alicorn?"

"She's my favorite alicorn. I've even considered sneaking back home and taking her. I would do it, but I'm having a hard enough time knowing what to do with Zebra."

"I wish I had an alicorn," Wren said wistfully. "If we hadn't left home, I would have saved up enough for one by now."

Tal raised one eyebrow. "Why doesn't your dad get you one? He has lots of money."

Wren shrugged. "My dad doesn't like to give me big things. He says I'll appreciate it more if I earn it myself."

"That's interesting. My dad gives me whatever I want, but he doesn't care about me. I'm going to have to think about what that means. What about you Li? Does your mom give you what you want?"

Ming Li put her knees up and wrapped her arms around them. "My mom's in the middle. We don't have the money the two of you do … although we have been getting a lot more now that my mom's baking is so popular. She gets me what I want for my birthday and holidays and stuff, but the rest of the year she makes me earn anything big. She buys all my clothes and essentials."

"Are you still going to take your mom to Meegore?" Wren asked, sitting on a chair.

"I forgot I didn't tell you about that. When I teleported home to grab some things, I talked to her. My mom is so stubborn. She says she won't go, and she doesn't want magic. I'm not giving up, though. If she went through, she'd be able to understand and speak the language better. It would make her life so much easier. I can't picture her climbing a wall or anything, though."

"Done," Tal said, handing the brush to Ming Li.

She ran her fingers through her long, black hair and they combed through smoothly. "Wow. That was impressive."

"Yes, I'm a man of many skills."

Wren fought a grin. "Man?"

Ming Li snorted. "More like a man-child."

"You two are impossible sometimes, but I'll assume what you meant was, thank you."

Ming Li grinned. "Yes, thank you."

Wren scooted back further in her chair and got comfortable. "So, are you going to tell us about Jaaz and Solia?"

Tal's eyes gleamed. "I thought you would never ask!"

Graham stood in the jungle, concentrating on making heat come out of his hands. "I don't even feel warm inside. Wren and Tal said they feel warm when they try it."

"I can't figure out why you can't do it," Sen said, scratching his head. "It seemed like one of the easier things to learn. Dovin is a superb teacher, though. Maybe I can't explain it as well."

"It must be me. I'm the only one not making progress."

"It could be because of the ice."

"What do you mean?"

Sen sat on a rock and looked at Graham. "Dovin said making warmth was one of the easiest things he could teach. He never had anyone who couldn't learn it. No one has been able to shoot ice as far back as anyone can remember. Maybe you can't make both cold and hot."

"I can see the sense in that. I can warm the ice into water though, but it doesn't make the water warm. It's still odd to think that Dovin taught you all these things. I wonder what his motive was."

"I don't know. He always seemed decent. Well, besides the fact that he ignored all the non-magical kids. I still find it hard to believe he's bad. He even told me he refused to teach Governor Briggs to teleport. That's what I should teach you. It would be more useful. I should teach everyone. It would save a lot of back and forth for Wren and I, and you wouldn't have to find one of us if you wanted to go somewhere."

"That sounds like a good idea. Is it hard?"

"It's hard to learn, but once you can do it, it's pretty easy."

"I can usually learn things pretty fast. That's why I'm surprised I can't make the heat."

"According to Dovin, some magic was lost because people didn't want to appear lazy. It wasn't that it was too hard, it was just starting to be socially unacceptable to do things the easy way."

Graham shook his head. "It's still surprising they would let something so useful disappear."

Sen stood and brushed off his pants. "I agree. We need to bring it back. Dovin did say his family worked on making some magic easier. Maybe teleporting used to be harder. Let's try now. Okay, hold on to me." Graham grabbed a piece of Sen's cape. "Now I'm going to think of where I want to go. I'll take us over there by that tree. Now I concentrate. The only way I can think of explaining it is you shut or fold in on yourself. It causes you to disappear, and as soon as everything feels dark, you open yourself up where you want to go. Are you ready?"

"Yes."

"Okay, here we go."

Graham focused on closing himself, and everything went dark. The times Sen had taken him somewhere, everything had spun, but this felt different. He was relieved Dovin hadn't taught Governor Briggs to teleport. It was good he had limitations. It was staying dark for too long. What had Sen said? Something about opening yourself back up.

He imagined himself opening up and landed in a brightly lit hallway. Where was he? And where was Sen? Something felt familiar, but he wasn't sure why. There were mirrors going down the hallway. He felt like he had been here before.

"Well, this is a surprise," a voice said behind him. Graham spun around and found himself face to face with Governor Briggs. "To what do I owe this honor?"

"Um ..." Graham racked his mind trying to figure out what to do. He shouldn't have let his mind wander. He hoped Sen was alright. "Wrong turn."

Briggs stoked his brown goatee and smiled. "How unfortunate."

Graham should probably try to teleport, but it hadn't gone smoothly the first time.

"I'll be going," he said, trying to walk past the governor.

Briggs stepped in front of him. "It would be rude to leave without catching up. Tell me where Tal is."

"I'm not sure. I haven't seen him today."

"But you know how to find him."

"Sometimes."

"It's rather heartless to withhold information about my son. I'm only worried about his well being."

"You don't need to worry. He's fine."

Briggs narrowed his eyes. "I need to talk to him."

"I'll let him know next time I see him."

"I've been told about the events on Meegore yesterday. You and your friends have taken on more than you can handle."

Graham shrugged. "We're all fine."

"The Dark Cloud is something for me to deal with, not children."

Graham crossed his arms. "Well, how are you dealing with them?"

"I will not share my plans."

"I thought you might not have any."

Briggs took a step towards him and stared Graham in the eye. "You are meddling where you are not needed. The Silver Eclipse all need to come to Akkron and follow my instructions. I am the law here and you *will* submit to me."

"I don't see that happening."

"I have a vision, and you are all part of it."

"Tal told us your vision. We aren't interested in being part of your kingdom. You're almost as bad as The Dark Cloud. What makes you think we would join with someone who falsely imprisoned Drew?"

"Drew," he said, rubbing his temples and looking at the ceiling. "That man has become a thorn in my side without even being here. He is a lot more popular than I believed. I suppose you don't want to tell me where he is?"

"Isn't he in your prison?"

"You know as well as I do he isn't. I don't know how he escaped. You know I can't have people taking the law into their own hands. If you would all come back to Akkron, we could work together legally. There should only be two groups and you are causing a third. That makes us weaker."

"What two groups should there be?"

Briggs glared. "My group and The Dark Cloud. Your little Silver Eclipse group is causing trouble. We're both fighting The Dark Cloud, so why aren't we doing it together?"

*"Where are you?"* Sen's voice asked, filling his head. Sen must know how to communicate better than the rest of them. He shouldn't be able to send a message without knowing Graham's location.

Graham ignored him. "Because your ideas are self-serving," Graham said, stepping back. "You want to be king? What's up with that? We aren't trying to help save Akkron from one evil takeover to hand it to another."

"I don't understand why you can't see this," Briggs said, shaking his head. "The difference is The Dark Cloud is evil and I am not."

Graham shook his head. "I guess it's relative. We don't feel like it's in the best interests of Akkron to have you all powerful. You seem more interested in your own power than the good of the people."

"I am what's good for the people!"

"Who are you talking to?" Valeena asked, coming around the corner. "Oh. Hello." She frowned when she saw Graham. Her hair was pulled up in a bun, with ringlets cascading down the front and she was wearing a fancy ball gown. They must have been at a party.

"Valeena, this is Graham. He is Tal's friend. He's being a little stubborn about telling me where our son is."

"And how did you come to be here?" she asked, pulling out a fan and waving it in front of her face.

"It was an accident."

Her lip turned down. "Strange accident. Are you here alone?"

"Yes. My friend should be with me, but he isn't." Graham wasn't sure how much to say in front of Briggs.

"Everyone is waiting for you downstairs," Valeena told her husband. "Why don't you go back to the party and I'll talk to Graham?"

Briggs frowned and glanced from Graham to Valeena. "Alright, but I want to talk to him later." He turned and stomped around the corner. Valeena grabbed Graham's arm and pulled him further down the hallway and into a dark room. He couldn't make anything out except a few shadows. She flipped on a light. It was some type of storage area, full of cleaning supplies.

"Why would you come here?" she hissed. "It's only slightly safer than being with The Dark Cloud."

"Sen was trying to teach me to teleport and I think I let my mind wander."

"And he isn't with you?"

"No. I was holding his cape when we started, but I don't know what happened."

She bit her lower lip. "I hope he didn't get lost or hurt."

"He must be fine. He sent me a message."

"And everyone made it through Meegore safely? We've been told a lot of strange things about that."

"Yes. We captured a few of The Dark Cloud. Austra was hurt, but she's alright."

"And what of Boztoll? We heard something went on there as well. I wanted to hurry back, but I needed to stay here."

"Valeena?" Briggs asked, startling them from the doorway. His eyes narrowed as he studied them. "What does this mean? You are conspiring against me?"

Valeena gasped and took a step deeper into the room. "No! Of course not."

"Then why would you be going to Boztoll?"

"I don't see how going to Boztoll would mean I was conspiring against you."

"You seem familiar with this boy. How could that be?"

Graham was surprised at the emotion in Brigg's eyes. He actually appeared more hurt than angry.

"I'm sorry, Briggs. You and I do not stand for the same things. I have to do what I think is right." She grabbed Graham's shoulder, and he felt himself spin away. They landed in front of Valeena's house in Boztoll.

Valeena's eyes looked tired. "Where were you when you lost Sen?"

"In the jungle, by the meetinghouse."

"I'll go there and check. You look here. If he was tele-porting somewhere, there is a good chance he's here or there."

"I'm sorry I messed things up." Graham had probably blown the best cover they had when it came to the gover-nor.

"It's alright," she smiled slightly. "I guess I don't have to fake it anymore. It is an enormous waste of time hosting parties and going to events when I could use my time more wisely."

"But you got inside information about the governor."

"Yes, well, I don't think his plans have been changing much. He comes up with a goal and sticks to it to an annoying degree. I'm going to go now. I'm worried about Sen." She disappeared and left Graham at the door. As soon as she disappeared Graham realized he could have sent a message to Sen. Too late now.

Graham heard someone yell his name. He turned and saw Wren running down the road towards him. Her red hair was flying behind her. He stepped away from the house and waited. She wasn't slowing. She plowed into him and wrapped him in a hug.

"You're alright!" her muffled voice said into his chest.

"I'm fine," he said, hugging her back. He glanced around to make sure Drew wasn't anywhere in sight.

She peered up at him. "Sen came a few minutes ago and said he tried to help you teleport and you disappeared! You didn't answer his message. Everyone's searching for you."

"I'm glad Sen's okay. I was careless."

She pulled back. "Let me send everyone a message." She put her fingers on her temples and closed her eyes. Graham knitted his eyebrows together. What was she doing? "Okay I told Tal, Sen, and Ming Li. All the adults that are searching for you are with one of them."

"Um, what did you just do?" Graham asked.

"Oh! I forgot to tell you guys. When we came out of the cave together, it made it so we could talk to each other through our minds. I'll explain more when everyone comes. There's Tal and Brake."

Graham looked  where she was pointing. Sen and Valeena appeared beside them, and Ming Li and Zera dashed around a corner.

"You cannot scare us like that!" Zera said, hugging him.

"Sorry."

"I thought you understood I was taking us somewhere. You were only supposed to observe," Sen said. "Why didn't you answer me?"

"Where did you go?" Brake asked.

Graham kicked at the ground. "I ended up at Governor Briggs' house. That's why I didn't answer. I didn't want Briggs to know what I was saying."

"Why go there?" Tal asked. "That seems like one of the last places you would want to be."

"I didn't mean to. I guess you can't let your mind get distracted when you teleport."

"Nope. That's one of the first rules," Sen said.

Tal looked from his mom to Graham. "Did my dad see you?"

"Yes. And worse than that, I blew Valeena's secret."

Valeena touched his shoulder. "Don't worry about that. I'm somewhat relieved, actually. It isn't easy to live with too many secrets."

Brake glanced around at all of them. "It's time to take a break from trying to make the silver eclipse."

"What?" Wren asked. "Why?"

"There are some things we need to do first. We all need to learn how to teleport or make a portal. It's a big hang up when only two people can do it."

Sen nodded. "That's what I was telling Graham. Although I probably should have explained, you need to clear your mind and not be distracted."

Wren tilted her head. "I'm not sure I can teach anyone to open a portal. I don't even remember the first time I did it. It would be better to have Sen teach teleportation."

"It's not too hard. I can teach everyone. The only thing I see that isn't as helpful as making a portal is that you can only teleport on this world. At least, that's all I can do. The Dark Cloud seems able to go to Earth whenever they want. This is where we need to be most of the time anyway, so that's probably fine. Valeena knows how to teleport so she can help."

Valeena winked at Sen. "I'd love to, but I should change out of this dress first. It isn't convenient for anything."

After a quick meal, everyone met in the forest around Boztoll and split into two groups. One went with Valeena and the other with Sen. Making a portal seemed as good as teleporting, so Wren didn't feel like she needed to join either group. Everyone was here now, even Brog. Giants had their own magic, but Brake didn't want him to feel left out.

Walking through the trees with nowhere to go felt refreshing. It had been a long time since Wren had had time to think. Now that she had the time, she wasn't sure what she wanted to think about. Her mind was tired. If they were going to take a break, they should definitely sleep more. Sometimes it was hard to be doing things all the time.

Something in Wren's head wanted to think on Graham, but that seemed like a silly thing to do when there were more important things that needed her attention. When Sen had come running in today to tell them Graham had disappeared, she had been terrified. She was glad that it had only lasted a few minutes.

Up ahead, Wren saw someone sitting on a log with their back towards her and their head down. She slowed and walked quietly. As she came closer, she recognized Tal's wavy brown hair.

"Hey," she said, trying not to scare him.

He jumped up quickly and spun around to face her. "You about gave me a heart attack!"

"Sorry."

He sat back on the log. "It's alright. I thought everyone was still practicing."

"They are. Why aren't you with them?" she asked, sinking down beside him.

Tal let out a breath and rested his elbows on his legs. "I'll wait until everyone learns it, then have Graham teach me."

Wren raised an eyebrow. "Why? It would be faster to get it done now."

"Probably, but I don't want to be in Sen or my mom's group."

"Why not?"

Tal gazed off into the trees. "If I say it out loud, it'll sound pathetic."

"It might help though."

"It will make you think less of me."

Wren put a hand on his arm. "It won't."

"My mom was never around much. Now that she's here and not as superficial as I thought, I feel like she cares more about Sen than me. She's been here making plans and trying to fix the world with Sen. I would have loved to have been with her all this time." He sighed and ran a hand through his hair. "See? That sounds lame. I'm jealous, pure and simple."

"That's not lame. I can understand how that might feel."

"If it ended there, it might be okay. I'm also jealous of the way Li seems so attached to him. There used to be a lot of back and forth between the two of us, but ever since Sen showed up, she only thinks about ways to compete with him."

Wren tilted her head as she studied him. "I didn't know you liked her that much."

"I don't. I mean, I like her, but I don't like her, like her. She's one of my best friends, but now she doesn't have time to talk to me or anything. And now Graham has parents. Don't get me wrong, I'm super happy for him, but now he spends time with them, which he totally should, but sometimes I feel a little out of the group. It should feel better, because now I have my mom, but at the same time, I don't."

"Have I been doing anything that adds to it? If I am, I want to know so I can change. I hope you know I'll always be your friend."

Tal gazed at her for a minute with a look she didn't understand. "I know. You're fine."

Wren wasn't sure what to take from that. "I'm fine? What does that mean?"

"I'm the only one that's the problem. These things don't usually bother me too much. I don't know why they are today." He stood and walked a few steps forward. "I should force myself to go join them."

Wren stood and walked to his side. She looked at him and frowned. "It's okay to have a bad day, but I can tell everything concerning me isn't fine. I want to know what I'm doing to add to your frustration."

Tal studied her with a lopsided grin. "I don't even know how to answer that."

Crossing her arms, she narrowed her eyes at him. "Well, you better try, because I'm going to keep bugging you until you tell me."

"It's not so much about you as it is my father."

"Your father? How does he have anything to do with me?"

"He doesn't directly."

"I feel so lost."

"My father has made some decisions for my life that I can't go against. In some ways it might be good, because it will keep me and Graham friends, but in some ways it stinks."

An alarm went off in Wren's head. It was telling her to offer her sympathy and leave. Still, Tal was her friend, and that made it her job to help him feel better, even if it made it awkward for her.

"I thought you were done listening to him."

"I am. But there are some things I can't change."

"Sure you can."

Tal lifted his hand and pushed a piece of her hair behind her ear. "I really can't." He dropped his hand and turned and walked away. Wren stared after him. Why were boys so complicated?

Graham watched Tal walk away from Wren, and he released a breath he had held a little too long. From where he was, it was impossible to hear what they had been saying, but he had almost been sure Tal was going to kiss her. Wren's eyebrows were knit together as she watched Tal disappear into the trees. At least Tal was as much of a failure as he was when it came to kissing a girl.

"Wren," Graham said, coming out of the trees. No point trying to avoid her.

"Hi," she said, shaking her head as if she were coming out of a trance.

"Don't you want to teleport?"

She shrugged. "Not really," she said, walking to meet him. "I'll go back with you though." They turned and walked towards Boztoll.

"Was that Tal here with you?" Graham asked, trying to sound bored.

"Yeah. He's having a bad day. I'm never good at helping people. I couldn't think of anything useful."

"What's bothering him?"

"The biggest thing is his mom and Sen."

"Oh. I wondered if that would bug him. She seems to pay a lot more attention to Sen than him."

"Yes. I should have told him to talk to her or something. She probably doesn't realize she's doing it."

"True. Valeena seems nice enough."

"He's upset about something his father did that he can't undo, but he didn't say what."

"Probably his marriage. I can't imagine anyone wanting an arranged marriage."

"Right ... I forgot about that. I never heard the details. Who is she?"

"He didn't tell me. He's never even met her."

Wren shook her head, "That sounds rough. I'm so glad my parents didn't do that to me."

"Yeah. I can't see why anyone would want one. Well, maybe someone like Hedder who won't get a girl any other way." Graham felt mean as soon as he said it.

Wren giggled. "Maybe you and Tal can find a girl for Hedder after you're done with Jaaz and Solia."

"That sounds terrible. I came out here to get away from Hedder. He learned to teleport super fast, and now he's looking down his nose at everyone else."

"But you already did it today."

"Yeah, but I did it wrong and my mom won't let me try again today. She thinks I need to learn to clear my mind better before I try again. I could totally do it, but I love having parents that tell me what to do."

"Didn't your aunt tell you what to do before?"

"Yes, but she did it out of meanness. They do it because they care. That makes an enormous difference. Can we change the subject before I make myself cry?"

Wren laughed. "Sure. I understand though. I can't tell you how many times I've wished my mom was around to tell me what to do."

Graham ran a hand over his hair. He wondered how hard it would be to change the arranged marriage law. It had to go deeper than just Akkron, because he heard people talking about arrangements between people in different cities. There was also the problem with what would happen if it did change. If the law were to disappear, he knew Tal would immediately pursue Wren. Now he was being selfish.

"Did you say something?" he asked, when he saw Wren looking at him funny.

"I was just wondering if you think things will work out with Jaaz and Solia."

"It's weird but I think they might. She told him they could be secret friends."

"Secret friends?" She shook her head. "I guess she hasn't changed much."

"Jaaz is a strange kid. He thinks one of her positive attributes is that she punched him in the nose and took his pencil."

"That sounds like Jaaz logic. Wasn't he the one bragging about me falling on his face?"

Graham laughed. "That's right. I'm hoping Solia isn't as bad as she seems. She takes her popularity too seriously, and it's taking a toll on her. Hopefully, Jaaz will help her realize there's more to life than what people think about you. I don't think he cares what people think."

"I can't even imagine what their conversations must be like."

"Ohhh, it's painful, I assure you."

# CHAPTER 19

"**I** think we should be doing something that isn't history," Ming Li said, slumping down at the kitchen table. "The world needs us, history doesn't."

"That's a pretty pathetic attitude," Professor Hedder said, glaring at Ming Li. He walked to his portable chalkboard and underlined the word history. "Can you think of no reason history might be important to you right now?"

"Nothing comes to mind," Ming Li said, shrugging.

"If we study history, it can help us with the future," Graham said from the head of the table. He could see a lot of good reasons for knowing history well. He hated agreeing with Hedder over Ming Li, though.

"Exactly," Professor Hedder said, nodding.

"But you keep teaching us about stuff that has nothing to do with what we're doing right now," Tal protested. "I can see the point in learning about The Dark Cloud, but you're teaching us about trolls. I don't see what they have to do with anything."

"The more you know about different species, the better you'll be prepared for something unexpected. If you know how they have acted through history, you can make a guess about how they might act if any conflict happens now."

Wren tapped her pencil on the table. "Do we know what the trolls and goblins will do if we end up in a war with The Dark Cloud?"

"It's hard to say," Hedder said, rubbing his sad excuse for facial hair. "The trolls are usually peaceful, and they usually choose the side that is doing what they think is right and ethical. The goblins are a whole different story. They usually go with whoever offers them the best deal. I've never known a goblin to do anything because it was right. They often fight for the good side, but only because of the compensation."

Graham thought of Meegore. "The giants are definitely on our side."

"Yes," Hedder agreed. "I've never known a giant that wasn't exactly what they appeared to be. It's too bad there are so few of them."

"Why is that?" Graham asked.

"How many female giants have you met?"

Graham thought for a minute. "Only Haltina."

"There are others besides her, but it's a problem. For the last hundred years or so, there have been ten male giants born for every one female. No one knows why, so it's a problem that can't be fixed. There is a rumor that there's a group of giants living on the other continent, but no one knows if it's true."

"I'm surprised no one has looked into that," Graham said thoughtfully. "Shouldn't Austra know?"

Hedder shrugged. "The rumor is about a *secret* group of giants, so the people there don't even know."

Ming Li shook her head. "Giants are huge. Would it be easy to hide a group of them?"

"The other continent has many uninhabited areas that nobody deals with. It isn't good farmland, so it isn't good for a lot."

"Wouldn't they need suitable land to grow things on?" Graham asked. "How are they getting food?"

"How do they get food on Meegore?" Wren asked, still tapping her pencil. "I didn't see any crops growing there."

"I can answer that," Brog said, barging through the kitchen door.

"Geez Brog!" Ming Li said, putting a hand to her heart, "Haven't you heard of knocking?"

"I am sorry," Brog said, bowing his head. "When I heard you talking about giants, I got excited. I have good hearing, remember?"

"I don't even know how giants get food," Professor Hedder admitted.

"Giants can grow anything, in any soil. All we do is put a seed into the earth and we sing. We do it at each meal. It grows quickly, and we are left with no waste. We only need a small piece of land."

"Why isn't this better known?" Professor Hedder asked, looking curious.

Tal tilted his head. "And why aren't you helping the people here who can't grow crops?"

"If people knew how easily we grow things, they might imprison us or try to take advantage. I only tell you, because I trust you. If we help every time humans have a problem, humans will stop trying to solve their problems and they will get lazier than they already are."

"That was harsh, Brog," Tal said with a gleam in his eye. "I'm wondering if giants are actually the lazy ones, though."

Brog crossed his muscular arms and glared at Tal. "How so?"

"Giants all live on a little island. You don't build houses, or buildings. You sing and your food grows? Sounds lazy to me."

"You do not get muscles like this from being lazy," Brog said, flexing his biceps.

Ming Li grinned. "I'm impressed."

Tal lounged back in his seat and smiled a crooked smile. "I don't know if being buff makes you productive."

"Are you trying to pick a fight with Brog?" Graham asked as he watched Brog clench his fists. "There isn't a chance of you winning."

"Nah," Tal said, "Brog just needs to get riled up now and then."

"You are teasing me?" Brog asked, his eyebrows knitted together.

"Yes, I'm teasing."

"And that is what humans do with their friends."

"Yes."

Brog nodded. "I think I am understanding. So you count me as a friend?"

"Of course we do."

Brog glanced around as Wren, Graham, and Ming Li nodded and a huge smile lit up his face, and slid back into a frown. "I was not teasing when I said humans are lazy. Are we still friends?"

Ming Li snorted. "Yes, we're still friends."

"A handful of powdered sunshine. It couldn't be that easy, could it?" Wren asked, as she handed the paper for the silver eclipse to Ming Li.

Ming Li looked at the paper. "As easy as what? We haven't made progress with any of this in over a week."

"What if the powdered sunshine is the dust from the cave?"

"I could see that. It's really powdery. Nothing like dirt."

"And it's bright for dust. If so, I have some in my cloak pocket. I never got around to cleaning it out."

"How would we know, though?"

"I don't know," Wren said, tapping her pencil against the table in the kitchen. "We could ask Padmire."

"I can't even tell you how much I don't want to go through the cave right now."

Wren nodded. She wasn't any more eager to go in than Ming Li. It had been an uneventful week, and going to the cave might change that. They had spent the entire week going to bed early, doing schoolwork, and learning to teleport and communicate with each other.

It was nice to talk to her friends from far away without getting a headache. Everyone in their group could now teleport, even Wren. Drew decided she should learn, just in case. When it became boring, she worked on her new levitating power. It was so much easier than before the cave.

"The cave might be easier now that we can both levitate. We could take ourselves up and across things. We wouldn't even have to get wet."

"Didn't someone say you couldn't levitate yourself through the whole thing or something like that?" Ming Li asked as she studied the paper.

"I don't remember. Did they?"

"I never listen, you know. I could have imagined it. Sen might have said it."

"We could send one of the boys. They could go through fast."

Ming Li shrugged. "Or we could open the cave and yell for Padmire."

"Or you could look down and see that he is already here," Padmire's squeaky voice said.

"Ahhhhhh!" Wren yelled.

"Geez!" Ming Li squealed, jumping off her chair. "Can't you knock or something?"

"Knock on what, your invisible door?"

"How did you get in?" Wren asked, as her heartbeat slowed back to normal.

"Brog brought me. I wasn't trampling around the world for you all again, especially with that enormous book. Now, why did you need me?"

"You brought *The Silver Eclipse* book?"

"Yes. You all seem determined to leave it behind or let it get destroyed. It is rather singed. So, why were you planning on coming to Meegore?"

Ming Li handed him the paper. "We were wondering if powdered sunshine is the yellow dust from the cave?"

"Of course it is. That's an easy one."

"Wren has some in her pocket, so that makes one ingredient that didn't take a lot of effort."

Wren cringed. "That's what you think. I had to go through all that stuff with Zalliah."

"Yes, but it's already done, so we can check it off," Ming Li smiled, making an invisible check mark in the air with her finger.

Wren pointed at the paper. "Do you know what these mean?"

"All the steps to make the silver eclipse?" Padmire asked. "Of course not. Remember, I can't read."

"Then how do you know we're right about the powdered sunshine?"

"Because that makes sense. Back in the old days, people always called it powdered sunshine."

"Alright. Another step done."

"Tell me another one. I might know it," he said, handing the paper to Wren.

"Water from a well of truth."

"A well of truth? None of your little group knows that one?" Padmire flashed his pointed teeth. "That's a fun one, to be sure."

Ming Li rolled her eyes. "So what is it?"

"About a thousand years ago, there was a man named Truth."

"Of course there was."

"Pretty clever, isn't it?"

"Annoying is more like it."

"Truth dug wells on the other continent. He was famous because his wells never went dry and the water never went bad."

"People don't use those kinds of wells anymore, do they?" Ming Li asked, glancing at Wren. "I've never seen any here."

"I don't think so. I've never been to the other continent though. Are any of the wells still there?" she asked, looking at Padmire.

"Why are you asking me? I've been locked away in the cave for a long time. Everything has changed. For everyone's sake, I hope they are still around."

Wren's mouth curved down. "Austra's from the other continent. We could see if she knows."

Ming Li smiled sympathetically. "You don't have to look so thrilled about it."

"I haven't talked to her since all the weirdness."

"You can't avoid her forever. The longer you do, the harder it will get. Besides, it's been over a week and I haven't seen your dad talk to her once. It could have been a one-time thing. Austra is picky and your dad is kinda weird. I don't see her settling on him in the long run."

Wren huffed. "He's not weird."

"I'm not saying a bad kinda weird, but if you think your dad's normal, I would say you're delusional."

"Where's Padmire?" Wren asked, looking under the table. Ben was the only one there, curled up on the floor, snoring.

"I don't think he sticks around for relationship conversations. He's probably searching for the others. Speaking of relationships, how are you and Graham doing? You haven't been tripping all over yourselves. Has anything happened?"

"We don't have a relationship, you know that."

"I don't trust you to tell me if that changes."

"I would tell you more things if you didn't blab to everyone."

"I don't see why everything is so secretive with you. It's like, three years ago when I told you that you had chocolate all over your lip. Most people would appreciate knowing, but you got all angry."

"That's because you yelled it from across the room! You could have come over and whispered or something."

"Well, I still think best friends have to share all relationship information."

"The closest to having a relationship in this group is you and Sen."

"Please," Ming Li said, rolling her eyes. "All Sen likes to do is compete. I'm fine with that, but I've realized you can't build a relationship with that. He sees me as someone who replaced his brothers. If we aren't competing, he doesn't even talk to me."

"Sen doesn't talk much to anyone."

"Well, it's annoying. I have things to say and I need someone who responds to let me know they're listening."

Wren tilted her head. "So you don't have a crush on him anymore?"

"I'm undecided. I might have liked him too fast because of that kiss. I'm pretty sure it was meaningless to him, so I wasted my time thinking about it."

"You know, you don't have to have a crush on someone. It's not like we're old or anything. We can all be friends and that's fine. It's probably best because a bad relationship might ruin everything."

Ming Li pushed her hair behind her back. "That's true, but I'm ninety-nine percent positive you will kiss Graham or Tal before this year ends."

"What? That's crazy. I'm not kissing anyone. And Tal has an arranged marriage, remember?"

"Yeah, but I can still see it happening. I'm guessing it will be Graham, though. I overheard your dad telling Hamble and Brog to make sure you aren't ever alone with Graham, so he must agree with me."

"What?" Wren exclaimed, fanning the flames from her face with her hand. "I can't believe him. Why are we talking about this? We should look for Austra."

"I didn't think you were eager for that."

"I'm not. The more I think about it, the more dread I feel, so that means I should talk to her as soon as possible and get it over with. It isn't worth losing sleep over."

"I don't see why you're nervous about it. All we need to talk to her about is the other continent. It's not like we have to ask her about the deep, hidden feeling of her heart. I'm sure she would hate that conversation as much, if not more, than you."

"Yes, but it seems weird. I haven't seen her much lately. I wonder where she is."

"Wonder where who is?" Austra asked, entering the room. Her blue heels clicked as she walked to the sink to get a glass of water.

"You actually." Ming Li said. "It's creepy the way everyone we talk about keeps appearing."

"You're looking for me?"

"Yes."

She sat at the table and glanced between the two of them. "Are you just wondering where I am, or do you need something?"

Wren chewed her bottom lip. Why was she so nervous? "We figured out the next thing we need to make the silver eclipse. It sounds like we might need to go to the other continent."

"Ah, I see. You finally need my help."

"Yes. Do you know anything about some wells that were made by a man named Truth?"

"There are quite a few stories about Truth. He helped the world in his time. He had a magnificent gift with his wells. Legend says that he placed enough around to help the entire continent. Most of them are gone now, but there are two we know of that still exist."

"Do they still have water?"

"Yes, but they aren't used by the public anymore."

"Who uses them?"

"There are two kingdoms on the other continent. There is one well in each kingdom. Can you guess who would have access to them?"

"The kings?" Wren guessed.

"Exactly," Austra said.

Ming Li slumped in her chair. "Does that mean we would have to talk to royalty if we wanted a cup of water?"

"That would be one way. It isn't easy to get an audience with anyone in the royal families. Sneaking in would also be pointless, as the wells are heavily guarded."

"We have to have the water for the silver eclipse," Wren said. "Do you have any ideas? It seems like the kings might be helpful if they know it could save everything. The Black Cloud could eventually spread."

Austra put her elbow on the table and rested her chin on her hand. She was quiet for what felt like far too long, but Wren let her think.

"My family used to be influential. My parents were wealthy and well respected. Everything changed when my mother died and my father married Melly's mother. I'm sure you remember Melly."

Wren shivered. "How could we forget?" Melly had tried to help Wren's uncle capture them not that long ago. She wanted to be part of The Dark Cloud, but had been captured by Tal before she could officially join them.

"Melly's mother didn't have any connections or money. She would spend more in a day than my parents ever had in months. My father didn't see any issues with the way she squandered his wealth. She wasn't very refined, and she didn't know how to speak civilly with people, so she burned a lot of bridges with our friends and family. It didn't take long before our family was a joke."

"Is that why you came to this continent?" Ming Li asked.

"Yes. Melly was born a year after my father married her mother. I tried to be a dutiful sister, but she was never interested in me. As soon as I was old enough, I left."

"So you can't help us?" Wren asked.

"I need to think about it. There might be some of my mother's friends that would be sympathetic to me. Do I have time to look into it?"

"It's our best shot, so I guess. Can you try to hurry, though?"

"Of course I'll try to hurry. You aren't the only ones who want this all fixed."

"So we need to go to the other continent?" Graham asked Brake as he mixed more berries into his cough elixir. Now that they could teleport, he was spending more time at his parent's house when he needed to work on something. It was a lot less distracting. His mom had moved back in, deciding she wasn't going to spend anymore of her life away from Brake.

Brake nodded and handed him a small bottle of honey. "That's what Padmire said. I wish I could go with you, but there are things here I need to take care of."

"What things?" he asked, putting a few drops of honey in the mix.

"We know you all have a lot going on, so we don't want to distract you with things we are doing."

"Who is we? All the adults in The Silver Eclipse?"

"Yes."

Graham mushed the berries some more. Nobody wanted chunky medicine. "But you all know what we're doing."

"We know The Dark Cloud is a priority, but we don't feel like that should be our focus right now."

"Then what is?"

"Governor Briggs."

"Oh," Graham said, looking sideways at his dad. "What are you doing?"

"We're gathering whatever evidence we can against him and we are going to have him voted out of office."

"I thought you couldn't do that."

"We are going to try, anyway. He won't allow an election, so we are waiting until he goes out of town and the Fleet can hold the election without him knowing."

Graham tasted the medicine and grimaced. He stirred in more honey. "Are all the Fleet on board?"

"We don't know yet. Zera thinks they will be. There is at least enough to convince the people to vote, and I don't think they will keep him in. He has fewer supporters than he did even a year ago. People want results and he isn't giving any."

"Who would replace him?"

Brake shrugged. "I believe anyone on the Fleet might run."

"Even Mom?"

"Doubtful. She has enough going on. She would do a great job, though."

"How hard would it be to get rid of arranged marriages? You know, if someone we knew was elected."

Brake's eyes sparkled. "Worried about those, are you?"

"Not for me. I know Tal isn't too thrilled about them."

"They aren't fair. I mean, it worked out for me, but Zera and I would have ended up together, anyway. Things might get more complicated for you if they were to get rid of them."

"Why?" Graham asked, sprinkling in a small amount of sugar.

"If Tal didn't have an arrangement, there's no telling where his interest would lie."

"Oh, I know where his interest would be," Graham mumbled. "I know where it already is."

"You know, I could talk to Drew–"

"No!" Graham said, as his heart sped up.

Brake laughed. "You know, Drew is already having you watched."

He dropped his spoon on the floor. "What?"

"Don't worry," he grinned, clapping Graham on the back. "Drew will be reasonable in the end."

"Everyone keeps acting like something happened between me and Wren, but nothing has. It's kind of strange talking to you about it." He picked his spoon off the floor and rinsed it in the sink.

"We missed enough of your life," Brake said. "We aren't going to miss teasing you about girls."

"What girls?" Zera asked, coming into the kitchen. "Are we talking about Wren?"

Graham shook his head. "Ugh! Why does everyone think I'm interested in Wren?"

"Oh sorry. Who were we talking about then?"

"Wren," Brake laughed. "Sorry kiddo. We all have eyes."

"I like Wren," Zera said.

Graham glared at them. "I thought we were focusing on getting rid of arranged marriages."

Zera nodded. "We have actually been discussing that in our Fleet meetings. Governor Briggs is against the idea of getting rid of them. I do not know who he has planned for Tal, but he believes it will be advantageous for their family. Even if Akkron abolished them, it would cause problems for those who already had contracts with people in other cities and on the other continent."

Brake flashed his white teeth. "I tried to tell Dryson that it might be in his best interest to let Tal keep his arranged marriage."

Zera playfully slapped Brake's shoulder. "Dryson doesn't want to marry Wren, only to find she was in love with his friend. It would be better to give Wren the choice than to have her settle."

"Marry? Come on, you two. I'm sixteen! I'm not thinking about getting married."

"Drew seems to be worried about what you are thinking," Zera said with a wink. "He told us you admitted to wanting to kiss Wren."

"I never said that! Drew kept asking me all these crazy questions."

Brake raised an eyebrow. "So you don't want to kiss her?"

"You two are enjoying this too much."

"We really are," Zera admitted. "I never thought I would get these opportunities." She went on her toes and kissed his cheek. "I need to go to a meeting. I hope you two have fun." She kissed Brake and left the kitchen.

"Whatever happens in the future, remember to choose someone who makes you happy. It's been so nice having Zera back. We were always happy. It was a long fifteen years, only seeing her when we could sneak moments here and there. It was a bad idea to stay apart. We should have helped each other through things instead of going along alone."

"Is it safe to let people know we're a family? Aren't you worried the governor will get suspicious?"

"The governor already knew Zera and I were married. He probably guessed you're our son. He hasn't bothered her at all, so he either isn't interested in us or he has some other plan. We can't live in fear. That's no way to do things."

"Isn't it strange he hasn't asked her where I am? He obviously wants to know where Tal is, and that seems like the most obvious way to find out."

"Perhaps he hasn't connected all the dots yet. He probably has so many schemes going on he can't keep them all straight."

"Is it safe for Mom to keep working with him though?"

"I don't know. We've discussed it, and she feels like it would set off more alarms if she stepped down."

"How's this?" Graham asked, handing his spoon to Brake.

Brake stuffed the medicine in his mouth. "I wouldn't call it tasty, but it didn't make me gag. It reminds me of something my mom used to give me when I was little."

"You probably have excellent control over your gag reflex."

"Are you saying that because I've choked down my cooking for so many years?"

Graham laughed. "I don't know why you didn't hire a cook or something. Some of your cooking brought tears to my eyes."

Brake grinned. "Oh, so now I'm your dad you can insult my cooking? I should have hired a cook. In the beginning, I thought we would find you faster and living by myself was only temporary. After that, I got used to living alone, eating my cooking, and I felt too lazy to change."

"It seems like you would get better after so many years."

"Nope. I didn't try hard. My mom tried to teach me when I was your age, but she gave up after I made everyone sick once or twice."

"So, Mom never made you cook?"

"I tried a few times when we were first married, and she made me promise to never do it again. She put me on permanent dish duty."

Graham poured the medicine into small vials. "I believe it."

"You shouldn't tease me. I've tasted a few of your medicines, and I would say the apple doesn't fall far from the tree."

"Mine are getting better, though. You have to admit that."

"It's true, and I'm proud of you. I have faith that you are going to change healing for the better. Some of those medicines you've made work impressively fast, like the one you used to heal that burn on Austra."

"It's exciting. I've always wanted to do this type of thing. It feels odd to not get to go to medical school, though. I have so much to learn and only books to help."

"You seem to get by. Hard work pays off. Now, why don't you show me how that basketball game works?" They walked out of the kitchen and headed for the ballroom.

"Hopefully, the new basket I put up lasts better than the one I made before."

"Next time you go to Earth, you should get a real one."

"That's a good idea."

"Alright, let's go."

"I have to warn you though," Graham smiled, "I'm good, and I'm not going easy on you because you're an old guy."

"Old guy? I still have a lot in me. I might surprise you. Zera and I were on a team that went to the championships in octaball."

"That was what, like fifty years ago?" Graham teased.

"Fifty years ago? You are lucky your mom isn't here to hear you say that. We were only twenty when you were born. If you're good at your math, you would realize that makes me thirty-six. I'm still in my prime."

"It sounds pretty ancient to me."

"I think you're stalling so you won't have to lose to me."

"Let's go see, shall we?" Graham smiled, as Brake laughed and put his arm on his shoulder. This was the way life should be.

"Hold still," Wren commanded, as she wrestled a squirming Ben into the bath water. "You can't stay in the meetinghouse with all these people if you smell like a goat." Ben hated baths, and thankfully dragons were clean animals so he didn't need one often. With everything going on, he was way overdue. He always fought it at first and slowly gave in and relaxed in the warm water. She would be happy if she stayed dry through the entire ordeal and even happier if her room didn't look like a hurricane had come through.

"Let me wash those wings."

"Can I come in?" Valeena asked from the doorway.

"Of course."

Valeena floated in and sat on Ming Li's bed. Wren rubbed soap onto Ben's wings and back.

"Can I ask you a question?"

"Sure."

Valeena twisted a handkerchief in her hands. "It's about Talon."

Wren glanced up and pushed a stray lock of hair out of her eye with her wet hands.

"I know I haven't been the greatest mom, so I've been giving him space. You seem to be good friends, so you probably know what he's thinking more than I do. It's important to heal our relationship, but I don't want him to feel pushed. Do you think he would react favorably if I were to make him some cookies or something? It's not my intention to manipulate him or anything."

"That would be a great idea," Wren said, pulling Ben out of the water and drying him with a towel. "He would like that."

"You think so?"

"I'm positive. He wants a good relationship with you. He doesn't want to ask for it because he fears rejection."

"Did he tell you that?"

"Not exactly, but I know he's a little jealous of the time you spend with Sen."

"Oh dear. I was trying to give him space. Thank you for telling me. I was worried about Sen now that his family is on Earth, so I tried to spend a bit of time with him while I was waiting for Tal to warm up to me. I guess I should have been trying with both of them."

"I don't think he's holding any grudges. He's looking for any signs you actually care to make him a bigger part of your life."

"Now I know what I need to do. Thanks Wren."

Wren nodded, and Valeena floated back out of the room. Wren rubbed some oil on Ben's wings so they wouldn't dry out. He licked her face, and she kissed the top of his head.

"Don't you feel better? You smell better."

Ben climbed onto her shoulders and curled around her neck like a scarf. Picking up the bin full of water, she carried it to the bathroom so she could dump it in the shower. She whistled as she cleaned it out. Hopefully Tal would be happier after this. He deserved it.

# CHAPTER 20

"What's that smell?" Graham asked, sniffing. The meetinghouse had never smelled this good before.

"I don't know, but I want whatever it is," Tal said, inhaling slowly. They walked down the hallway and into the kitchen.

"Tal! You're back!" Valeena said, smiling. "I made you some cookies." She grabbed a plate and offered it to him.

"You made them for me?" Tal asked, taking a cookie.

"Yes. I made them for you a few times when you were little and you loved them."

She offered one to Graham and put the plate on the table. Graham bit into it and closed his eyes. They were good.

Tal chewed slowly and smiled. "Wow, mom. These are fantastic."

She beamed. "Thanks. I always loved baking."

"Really?"

"Yes. I don't usually do it anymore because Briggs said that's what servants are for. I occasionally slip in and make something when he's out of town. You remember the cake we only have when he's gone?"

"You make those?"

"Yes."

Graham grabbed another cookie and excused himself. This seemed like a time for Tal and Valeena. There wasn't a lot to do in the meetinghouse, so he wandered outside for some fresh air. Having this week to rest and learn new things was nice, but Graham was ready to go back to work.

If they were going to travel to the other continent, they needed to talk to Austra. She was the only person he knew who was from that place. Graham heard a sound to his right, so he turned left. There weren't a lot of animals around here that he wanted to run into. Brog took Zebra back to Meegore against Tal's wishes, because a jungle wasn't a safe place for a pet Zebra. On Meegore, she could run free because there weren't any predators.

There were more noises than usual today. It was common to hear birds and monkeys, but Graham felt like he was hearing other things he couldn't identify. He was probably being more paranoid than normal. He heard another rustling sound and turned the other way. There weren't as many bugs now that The Dark Cloud had messed up the weather. It was warm and humid, but not as hot as it should be.

Graham felt uneasy, but he wasn't sure why. He felt like he was being watched, so he walked faster and tried to shake the feeling. He started running and told himself he wasn't scared, he just wanted to get out of the humidity, but he knew it wasn't true.

Tal appeared from behind a tree and Graham slowed down. "Graham! Something's wrong."

"What?" he asked, as Tal turned to walk with him.

"I don't know. Zera sent a message to my mom and said we need to leave the meetinghouse immediately. Wren and my mom are out looking for you, too. We tried to message you, but you were blocking us or something. Wren has Ben, so we don't have to go back for anything. I'll send them a message–"

"Well, imagine running into you two out here!" Zalliah said, stepping out in front of them. Graham jumped back.

Tal glared at the woman. "What do you want?"

"So many things," she said, pushing the hood of her cloak down. Her brown hair hung limp in the humid air. "Most of them I'm not sharing today. I could use a job, since I can't trust Wren to not ruin my cover. Hiding is such a silly thing. Speaking of cover, here come the clouds."

Graham and Tal gazed up through the trees at the sky and watched it darken.

"Why do you keep doing this?" Graham asked. "You are going to ruin the weather everywhere and that will affect you, eventually. Even evil people have to eat."

"Yes, and we are trying to do it sparingly, but we thought you were worth the trouble."

Tal shrugged. "We aren't scared of clouds."

"Is anyone really? Clouds aren't anything to fear. It's what happens when the clouds are there that is scary."

Graham and Tal glanced at each other. What was she stalling them from?

Zalliah raised one eyebrow. "No one is going to ask what the clouds are hiding today?"

"Probably a fire," Tal said. "That seems to be as creative as The Dark Cloud gets."

"Why fix what isn't broken?"

"How did you find us?" Graham demanded.

"Ah, that has been difficult. I may have put a bee in Governor Briggs' ear. The fight at Meegore was a failure, to say the least, but it let me know who the members of your group are. The governor was only too happy to join forces for some of that information."

Tal clenched his fist and Graham was pretty sure he heard his teeth grind together.

"Yes, poor Tal. It must be hard to have that man as a father. Such a disappointment."

"Wait," Graham said, "If the governor knows everyone in the group ..."

"Then he must know Zera can't be trusted," Zalliah smiled. "How sad for you. One less voice on the Fleet. I sure hope she isn't arrested at her meeting today."

Graham narrowed his eyes and grabbed Tal's arm. "Let's get out of here. She's distracting us from something."

Zalliah laughed as they turned and sprinted towards the meetinghouse. Graham could smell fire. They couldn't have found the invisible house, could they? Tal must have smelled it as well, because he ran faster. Graham hoped his mom was okay, but they had to make sure Wren and Valeena were safe first.

They rounded a corner and came to a stop at the sight ahead of them. The house was burning. Now that it was on fire, it wasn't invisible. Graham could tell immediately that the two story building couldn't be salvaged. Wren and Valeena ran to them, and they all turned to watch.

"How did they find it?" Wren asked, hugging Ben close to her. In her other hand, she had *The Silver Eclipse* book.

"We talked to Zalliah," Graham said. "I bet they had my mom followed. It sounds like the governor knows who everyone in our group is now. I should send my dad a message."

"I'll send a message," Valeena volunteered. "You put out that fire."

"I don't think we can save it."

"No, but we don't want the entire jungle to burn."

Graham nodded and moved closer to the fire. He raised his arms and shot ice towards the fire. It took a few minutes before he could turn the ice to water, but once it started, it was a steady stream. Graham wondered where all the ice and water was actually coming from. It couldn't all be coming from him or he would run out. Sometimes magic was amazing.

Sweat dripped down his forehead, and he thought he was seeing things when the nearby trees seemed to pull back from the fire. Out of the corner of his eye, he saw Tal waving his hands through the air. He must be moving the trees to give him more time.

The outside of the house wasn't painted. It made sense if no one was going to see it. It was pointless decorating something that was invisible. Graham kept the water coming, but he was feeling faint. It was anyone's guess how long he could keep this up.

"Let it go," Valeena said, putting a hand on his shoulder. "Tal moved the trees far enough. It should be safe."

Graham nodded and let his arms fall to his sides. He sank to his knees and watched the flames. The house wasn't huge, so it wouldn't take long. Wren and Tal helped him to his feet.

"It's too hot over here. Let's move back," Wren said, pulling on his arm. They moved away until the fire's heat was less intense.

"Brake wants everyone to meet at his house," Valeena told them.

"Look there," Tal pointed into the trees. The clouds made the jungle dark now, but there was a blue glow not too far away. He laughed. "Nice glow Zalliah!" he yelled. "I hope you like your gift from Meegore!"

Wren opened a portal, and they all scurried through.

Wren's eyes wanted to close. She scooted herself into the corner of Brake's sofa and laid her head on the armrest. The entire Silver Eclipse was gathered in Brake and Zera's sitting room. Brake brought in extra chairs so everyone would be comfortable. Everything had to change. If The Dark Cloud and Governor Briggs knew who they all were, then they all needed to go into hiding. Everyone worried the governor wouldn't let Zera go, but as soon as she realized he was onto her, she teleported home and sent them all the warning.

Two hours had passed, and nobody could compromise on the next course of action. All that had been agreed on was that Brog should stay at the prison to make sure the prisoners were all cared for. There were two people who stayed there full time, but having Brog there made everyone feel better.

"The longer we stay here, the more likely it is we will get caught," Hedder said. "Everyone knows where Brake lives, so someone will come here eventually."

"We could go to the cabin, but nobody likes to be there," Drew said.

"We have a place in Flordillia," Zera said, shifting on her chair, "But I do not know if the trolls would be happy to have a bunch of humans hiding there. It is the largest place we have."

"We could stay at Meegore in the cave," Tal suggested. "Nobody else could get in."

"Yes," Wren said, "But the crystal would glow so everyone would know we were there."

"Right."

"Why don't we all go to the other continent since the five of us have to go, anyway?" Ming Li suggested.

"We're too close to overthrowing the governor," Brake said, rubbing his chin. "The adults need to stay close to Akkron."

"It is going to be a lot harder now that I cannot talk to the other Fleet members," Zera said, tapping her fingers on the armrest. "I do feel like they would talk to me if I could get to them. I am fairly certain at least four of them will side with us."

Valeena nodded. "Anelle, Lyna, Mav, and Jayla would side against my husband. Petral, on the other hand, was meeting with Briggs secretly almost every night I was home. I don't know why, but I assume it can't be good."

"That is the way I was feeling as well. I have never trusted Petral, and he always sides with the governor."

Wren decided now was as good a time as any to bring up her thoughts on The Dark Cloud. "I know this has nothing to do with the governor, but I might have an idea about how to defeat Zalliah."

"Oh? How is that?" Brake asked.

"When I talked to Zalliah, and she told me she wants to destroy The Dark Cloud and make herself a hero, it gave me an idea. Obviously, some of the other members don't know her plan or they wouldn't be going along with it. What if we let them know? They might get angry and fight against her themselves."

"It's something to think about," Brake said, rubbing his temples. "My worry with a plan like this is that she might not have been telling you the truth. It would be careless of her to tell you a plan like that, because there would be nothing to stop you from telling the rest of the world. She also told you she would have anyone in on her plan kill you if you told them, and we aren't risking that."

"But she was planning to make me go with her, so she might have thought it was safe to tell me."

"True, but it seems too sloppy for a plan she's had in place for so many years. If it was her plan she might change it, now that you know what it is."

"Well, we should still tell the world, so if it is true, nobody will follow her."

Tal looked at the floor. "We should tell my dad."

"I don't think that's a good idea," Valeena said, shaking her head.

"Dad might be corrupt, but he doesn't want to end up getting thrown out of office and having Zalliah take over. If he knew she was part of The Dark Cloud, he might end any relationship he has with her and might work harder on getting rid of them."

Valeena sighed. "It's possible I guess."

"There is something else that needs to be discussed," Zera said, biting her lip. "We debated whether or not you five should be told. Some felt you should be sheltered from it."

"What is it?" Graham demanded. Wren braced herself. She wasn't sure she wanted to know.

"When we fought at Meegore, there were more casualties than you are aware of."

Wren raised her eyebrows. They knew one of The Dark Cloud members had died.

"We decided to tell you because you need to be careful. You need to realize this is all real and not just a game."

The corners of Tal's mouth turned down. "We know it's real. When have we acted like it's a game?"

"Not a game," Brake said leaning forward. "But sometimes you don't stop to think. We want you to be careful. There are consequences to poor planning."

Wren swallowed hard. Was the fight at Meegore her fault? She was the one who got taken there.

"We know that Meegore was nobody's fault. We just want you to all be extra careful."

"What other casualties?" Wren asked, not sure she wanted to know.

Zera looked into Wren's eyes. "Two more Dark Cloud members died as soon as they reached the prison, and one giant."

"A giant?" Wren muttered. "Which one?"

"He was not one that you know. Brake is right. No one is to blame except Zalliah and her followers. I don't want any of you to feel guilt over it."

The room was silent while they all wrestled with their thoughts. Zera was right. This wasn't on them. That didn't make it any less sad though.

"Aren't we supposed to be discussing where we should go?" Ming Li finally asked. "If we keep jumping around to other problems, we'll be here all night."

"The five of us need to go to the other continent," Graham said. "Austra could tell us about it."

"We already asked her," Ming Li said. "It sounds like the only wells that are still around that were built by Truth belong to the royal families."

Wren rubbed her eyes. "I thought the wells never went dry. That's what Padmire said."

"They were purposely destroyed," Austra explained. "It didn't all happen at once. It usually happened as a result of war."

"So what do we do?" Graham asked.

"I haven't had time to check with anyone to see if we can get any help," Austra said, dusting invisible lint off her blue dress.

"Well, which kingdom are we most likely to have success in?"

"Success by talking to the king, or success by breaking in and taking the water?" Tal asked.

Graham shrugged. "Either."

Austra looked thoughtfully up at the ceiling. "The Northern Kingdom might be the easiest, because it's always been the most corrupt."

"Why would that make it easier?" Wren asked. She couldn't imagine working with corrupt people as being a good plan.

"Making a deal with corrupt people is easier, because they don't have as many rules. Years ago, there was a king named Rihan. He was a horrible king. His council was full of the worst types of people you can imagine. He taxed the people to destitution. King Rihan and his council held elaborate parties and events while his people worked to support his lavish lifestyle.

"The people didn't revolt until one man finally took a stand. He told the king to his face that if he didn't stop exploiting the people, they would eventually rebel and he would be taken off the throne. The king didn't like that at all. He had the man put to death. That was where he made his mistake. The man was popular among the people, and so they did rebel. It wasn't hard to overthrow King Rihan, because the only people loyal to him were his council.

"They placed Rihan's son on the throne, under the condition that he not rule them like his father. King Miadd is a much better king than his father was, but he is still corrupt. He doesn't overtax the people, so they are okay with him. He might be easier to make a deal with than the king of the Southern Kingdom."

"What kind of deal can we make?" Wren asked. "I can't imagine anything we have being valued by a king."

"I'm not sure," Austra admitted. "We might have to think about that."

"I saw a light outside," Hamble said from the window. "It's time to go."

"We accomplished nothing," Hedder complained. "Where are we all supposed to go?"

"Tonight we can stay in Flordillia," Brake said, "And with luck the trolls won't know we were there."

"Does that mean we're just moving locations so we can argue all night?" Ming Li asked. "Because seriously, if that's what we're going to do, I need to be mentally prepared."

"We will sleep tonight and talk tomorrow. Alright?"

"Fine," Ming Li agreed, "But there better be beds at this place."

"Why weren't we staying at this place to begin with?" Ming Li asked as soon as they stepped out of the portal.

Graham had to agree. The house was enormous. They were standing in an entryway that had a double staircase curving up both sides. The walls were gray and white swirled marble, and between the two staircases was a granite fountain with two fish squirting water out of their mouths.

"Who takes care of this place?" Tal asked.

"I stay here most of the time," Hamble said. "I'll be glad to have company. It's pretty lonely by myself."

Wren leaned in and whispered to Graham. "After you find a girl for Hedder, find one for Hamble." Graham smiled, but didn't say anything. If his friends from Missouri knew he was matchmaking, they would laugh for days.

"There are around thirty bedrooms," Hamble told them. "All upstairs. There are bathrooms hooked to most of them."

"Really? Why haven't we been staying here?" Ming Li asked again.

"Trolls dislike living around people," Zera said. "The more people, the more nervous they get."

Graham gazed up. The ceiling seemed so high. "How did you even get a place like this?"

"Trolls are extremely superstitious," Hamble said. "The troll that lived here liked to experiment with different elements and one day he blew himself up. He had a lot of family, but none of them wanted to live in a place where someone had been blown up. They were happy to let us take it. We paid, of course, but not as much as it was worth. They wanted us to take it for free because they thought any money would also be tainted."

"I learned to speak Flordillian and I've never met a troll, so I couldn't use it," Ming Li said. "Maybe I'll get my chance."

"Quick question," Tal said, looking up the stairs. "What are the rules about sliding down the rails?"

Austra rolled her eyes. "Do we need to discuss this?"

"No rule," Hamble said. "Drew, Brake, and I have had some good races in our time."

"Everyone needs to sleep," Drew said, yawning. "Things will make more sense in the morning."

"Everyone go find a room," Zera said. "The ones that we use are locked, so take any of the others."

"Race you for the best room!" Sen shot up the steps. Ming Li was close on his heels.

"If it has thirty rooms, I'm not running," Tal said, walking slowly up the steps. Wren and Graham followed him. At the top of the stairs, there were three hallways. They could go left, right, or straight.

"We should probably choose rooms close together," Graham said. "Just in case." They followed Sen and Ming Li. The hall was long, with lots of doors.

Tal opened the first door they came to and looked in. "This is fine. See you guys tomorrow."

"I wonder what a troll city is like," Wren said, opening the next door. "This room has a balcony. Do you want to look?"

"Sure," Graham said. They walked into the room and glanced around. "This room is kinda spooky." The bed was immense, with a fluffy red bedspread and at least six giant pillows. Across from it was a suit of armor, much too big for a person. Above the bed was a painting of what must be a troll. It was thick with huge muscles. Its hair resembled porcupine quills, and it had a metal breastplate and leather pants. He held a sword in his hands and his mouth was twisted in a snarl that showed his sharp yellow teeth.

"You can have this room," Wren said. "That guy would give me nightmares."

"I was picturing trolls as cute little creatures with blue and pink hair."

"That seems kind of random. I've never heard of a cute troll. And pink and blue hair? Let's look outside."

They opened the balcony door and gazed out at the city. The trolls had skyscrapers. They weren't huge, but they were the first ones he had seen since leaving Earth. The moon was full and lit up the city nicely.

"Wow, those buildings are so tall!" Wren said, looking around. "I thought trolls would live in dirt houses or something like that."

"That's what I thought from all the history books," Graham agreed. "I guess history doesn't talk about modern things."

"I'm surprised my dad never told me about this."

"It's pretty clean. I would have guessed from the painting of that troll that they wouldn't be bothered with cleaning."

"It's probably a warrior picture, although trolls stay out of wars if they can help it. I'm hoping there aren't pictures like that in every room."

"If there are, we can put them in the hallway."

Graham leaned his forearms on the balcony railing, and Wren imitated him. "Are you ready to take on the other continent?"

"You sound like we're going to war with them."

"I hope not. Hopefully, it's an in and out kind of deal."

Wren gazed at the moon. Her red hair glowed in the moonlight. "It's nice to see the moon not covered in clouds. It seems huge here. I love how peaceful it is. I wish that feeling could stay."

"We're getting closer to creating the silver eclipse, and the adults are getting closer to taking down the governor. It shouldn't be much longer and everything will be more peaceful."

Wren peered at him with her sparkling emerald eyes. Her green shirt made them stand out even more than usual. "That will be nice. I can't wait. What will we do after that? Go back to school and act like everything is normal again? Somehow, that doesn't sound possible."

"I don't know. It does sound weird." The thought crossed Graham's mind to tell Wren her shirt was green. He smiled out into the darkness. It kinda worked for Jaaz.

"Why are you smiling?" she asked.

"Oh, I was thinking about Jaaz."

"Oh," Wren said, studying the buildings. Did she seem disappointed? Graham wasn't sure, but it looked possible. Did she want him to be thinking about her? Love was so complicated. Wait, did he just think love? He meant girls. Girls were so complicated.

"Jaaz asked me if I ever told you that you have beautiful eyes," Graham said, forcing himself to look at her.

"What?" Wren asked, blinking.

"You know how Jaaz is. He told me you have beautiful eyes and that I should tell you."

"Interesting," Wren said, studying her hands.

"He's right. You do have beautiful eyes."

"Thanks," she said, quietly.

"And just so you know, if the troll picture comes alive at night, I will totally fight it for you."

Wren smiled and gazed up with a twinkle in her eye. "And I will totally fight it for you."

"Great. Between the two of us, it doesn't stand a chance."

"And neither does The Dark Cloud."

"No chance at all. We've got this."

"We do."

# ACKNOWLEDGEMENTS

Thank you for reading my book! I hope you enjoyed it. If you did I would appreciate a review. Without reviews it is hard to make a book visible in this competitive market.

A special thanks to my family and friends who have been supportive and helped me through the writing process. There are so many that I can't name them all. I especially need to thank my husband and kids who put up with me when I'm in the writing zone.

# About the Author

Kristy Dixon graduated with a degree in English from the University of Utah. She started writing stories when she was seven and has been writing ever since. In the beginning she tried writing historical fiction, but soon found that to be boring. She started writing The Silver Eclipse at the urging of her children and found fantasy writing to be more fun. She has nine children and six chickens. If she isn't writing or playing board games with her kids, she is probably eating cookies or wishing that she were eating cookies. You can contact Kristy at kristydixon35@gmail.com and find her website at kristydixonbooks.com.

# ALSO BY KRISTY DIXON

The Silver Eclipse: Akkron

**Coming Soon**

The Silver Eclipse: The Other Continent